Bainite and Martensite

Bainite and Martensite: Developments and Challenges

Special Issue Editor

Carlos Garcia-Mateo

MDPI • Basel • Beijing • Wuhan • Barcelona • Belgrade • Manchester • Tokyo • Cluj • Tianjin

Special Issue Editor
Carlos Garcia-Mateo
National Center for
Metallurgical Research
Spain

Editorial Office
MDPI
St. Alban-Anlage 66
4052 Basel, Switzerland

This is a reprint of articles from the Special Issue published online in the open access journal *Metals* (ISSN 2075-4701) (available at: https://www.mdpi.com/journal/metals/special_issues/bainite_and_martensite).

For citation purposes, cite each article independently as indicated on the article page online and as indicated below:

LastName, A.A.; LastName, B.B.; LastName, C.C. Article Title. *Journal Name* **Year**, *Article Number*, Page Range.

ISBN 978-3-03928-857-1 (Hbk)
ISBN 978-3-03928-858-8 (PDF)

Cover image courtesy of Carlos Garcia-Mateo.

Contents

About the Special Issue Editor

Carlos Garcia-Mateo is a physicist specializing in solid state matter. He obtained his Ph.D. from CEIT-TECNUM in San Sebastian under the supervision of Prof. Rodriguez-Ibabe. He later joined the group of Prof. Bhadeshia at Cambridge University (2000–2003) to work on nanostructured bainite, his main area of research interest to date.

His research has received the international recognition through the following awards:

- Meritorious Award for Best Products and Forging Paper on 26 October 1999 at the 41st Mechanical Working and Steel Processing Conference in Baltimore, Maryland.

- The Vanadium Award-Council of the Institute of Materials, Minerals and Mining-UK, sponsored and selected by the Vanadium International Technical Committee (Vanitec), for the most outstanding paper in the metallurgy and technology of vanadium and its alloys, in the years 2000 and 2008.

- The Cook/Ablett Award 2015-Council of the Institute of Materials, Minerals and Mining-UK for work of particular merit in the field of metals published in one of the Institute's journals.

He has published more than 120 papers in SCI journals, with an h-index of 35 representing a total of more than 3000 citations. He has co-authored 12 chapters and supervised numerous M. Phil and Ph.D. students, mostly in the area of bainitic steels.

Editorial

Bainite and Martensite: Developments and Challenges

Carlos Garcia-Mateo

MATERALIA Research Group, National Center for Metallurgical Research, CENIM-CSIC, Avda, Gregorio del Amo 8, E-28040 Madrid, Spain; cgm@cenim.csic.es; Tel.: +34-91-5538900

Received: 14 November 2018; Accepted: 17 November 2018; Published: 19 November 2018

1. Introduction

Both microstructures, martensite and bainite, although share some common features, when studied in further detail depict a plethora of subtle differences that make them unique. Tailoring the final properties of a microstructure based on one or the other as well as in combination with others, exploring more sophisticated concepts such as Q & P and nanostructured bainite are the topics of worldwide research. Understanding the key microstructural parameters controlling the final properties, as well as the definition of adequate process parameters to attain the desired microstructures, goes undoubtedly through a proper understanding of the mechanism ruling their transformation and a detailed characterization.

The development of new and powerful scientific techniques and equipment (EBSD, APT, HRTEM) allow us to gain fundamental insights that help to establish some of the principles by which those microstructures are known. The developments accompanying such findings lead to further developments and intensive research providing the required metallurgical support.

2. Contributions

The present Special Issue includes one review paper [1], one technical note [2] and ten scientific papers [3–12]. In all of them, martensite and/or bainite are being studied and in some cases, in combination with other phases.

New processing routes by plastic deformation of austenite previous to transformation, ausforming, is revealed as an emerging and promising alternative to achieve optimized microstructures [1,4]. Development of new steel grades for specific industrial products and existing processing routes [11]. The response of martensitic and bainitic microstructures to more traditional treatments as nitrocarburising [9] or tempering [2,3,5] based on the initial microstructural characteristics. More fundamental studies on the ways and means of phase transformation in specifically designed alloys [6,8,12] and the mechanical response and relationships between microstructure and mechanical properties [2,7,10] are among the topics presented in this compendium.

3. Conclusions and Outlook

Regardless of the sector, the driving force that moves and promotes development in the field of materials is the search for better properties at a lower cost and with less environmental impact. Some of us are lucky enough to work with one of the cheapest, most versatile and oldest materials available, steel. Even to this day, there exists in certain circles, the belief that everything in the world of steel is already investigated. The works presented here are not more than the tip of an iceberg that come to demonstrate the little foundation of such affirmations and the good technical and scientific health that this material enjoys.

As a guest editor, I would like to express my sincere thanks to all my colleagues for supporting this initiative and share their latest developments, making this Special Issue a total success.

Conflicts of Interest: The author declares no conflict of interest.

References

1. Garcia-Mateo, C.; Paul, G.; Somani, M.; Porter, D.; Bracke, L.; Latz, A.; Garcia De Andres, C.; Caballero, F. Transferring Nanoscale Bainite Concept to Lower C Contents: A Perspective. *Metals* **2017**, *7*, 159. [CrossRef]
2. Sourmail, T.; Garcia-Mateo, C.; Caballero, F.; Morales-Rivas, L.; Rementeria, R.; Kuntz, M. Tensile Ductility of Nanostructured Bainitic Steels: Influence of Retained Austenite Stability. *Metals* **2017**, *7*, 31. [CrossRef]
3. Yen, H.-W.; Chiang, M.-H.; Lin, Y.-C.; Chen, D.; Huang, C.-Y.; Lin, H.-C. High-Temperature Tempered Martensite Embrittlement in Quenched-and-Tempered Offshore Steels. *Metals* **2017**, *7*, 253. [CrossRef]
4. Vivas, J.; Capdevila, C.; Jimenez, J.; Benito-Alfonso, M.; San-Martin, D. Effect of Ausforming Temperature on the Microstructure of G91 Steel. *Metals* **2017**, *7*, 236. [CrossRef]
5. Talebi, S.; Ghasemi-Nanesa, H.; Jahazi, M.; Melkonyan, H. In Situ Study of Phase Transformations during Non-Isothermal Tempering of Bainitic and Martensitic Microstructures. *Metals* **2017**, *7*, 346. [CrossRef]
6. Luo, Q.; Kitchen, M.; Abubakri, S. Effect of Austempering Time on the Microstructure and Carbon Partitioning of Ultrahigh Strength Steel 56NiCrMoV7. *Metals* **2017**, *7*, 258. [CrossRef]
7. Larzabal, G.; Isasti, N.; Rodriguez-Ibabe, J.; Uranga, P. Evaluating Strengthening and Impact Toughness Mechanisms for Ferritic and Bainitic Microstructures in Nb, Nb-Mo and Ti-Mo Microalloyed Steels. *Metals* **2017**, *7*, 65. [CrossRef]
8. Grajcar, A.; Zalecki, W.; Burian, W.; Kozłowska, A. Phase Equilibrium and Austenite Decomposition in Advanced High-Strength Medium-Mn Bainitic Steels. *Metals* **2016**, *6*, 248. [CrossRef]
9. Fabijanic, D.; Timokhina, I.; Beladi, H.; Hodgson, P. The Nitrocarburising Response of Low Temperature Bainite Steel. *Metals* **2017**, *7*, 234. [CrossRef]
10. Celada-Casero, C.; Kooiker, H.; Groen, M.; Post, J.; San-Martin, D. In-Situ Investigation of Strain-Induced Martensitic Transformation Kinetics in an Austenitic Stainless Steel by Inductive Measurements. *Metals* **2017**, *7*, 271. [CrossRef]
11. Bracke, L.; De Knijf, D.; Gerritsen, C.; Hojjati Talemi, R.; Diaz Gonzalez, E. Development of Direct Quenched Hot Rolled Martensitic Strip Steels. *Metals* **2017**, *7*, 326. [CrossRef]
12. Allain, S.; Geandier, G.; Hell, J.-C.; Soler, M.; Danoix, F.; Gouné, M. Effects of Q&P Processing Conditions on Austenite Carbon Enrichment Studied by In Situ High-Energy X-ray Diffraction Experiments. *Metals* **2017**, *7*, 232.

 *metals*

Article

Effect of Ausforming Temperature on the Microstructure of G91 Steel

Javier Vivas *, Carlos Capdevila, José Antonio Jimenez, Miguel Benito-Alfonso and David San-Martin

Centro Nacional de Investigaciones Metalúrgicas (CENIM), Consejo Superior Investigaciones Científicas (CSIC), Avda Gregorio del Amo, 8, E 28040 Madrid, Spain; ccm@cenim.csic.es (C.C.); jimenez@cenim.csic.es (J.A.J.); mba@cenim.csic.es (M.B.-A.); dsm@cenim.csic.es (D.S.-M.)
* Correspondence: jvm@cenim.csic.es; Tel.: +34-91-553-89-00; Fax: +34-91-534-74-25

Received: 18 May 2017; Accepted: 23 June 2017; Published: 27 June 2017

Abstract: The development of thermomechanical treatments (TMT) has a high potential for improving creep-strength in 9Cr-1Mo ferritic/martensitic steel (ASTM T/P91) to operate at temperatures beyond 600 °C. To maximize the number of nanoscale MX precipitates, an ausforming procedure has been used to increase the number of nucleation sites for precipitation inside the martensite lath. Relative to standard heat treatments (consisting of austenitization at about 1040 °C followed by tempering at about 730 °C) this processing concept has enabled achieving a microstructure containing approximately three orders of magnitude higher number density of MX precipitates having a size around four times smaller in ASTM T/P91 steel. On the other hand; this TMT has little effect on the size and number density of $M_{23}C_6$ particles. The optimized microstructure produced by this TMT route proposed is expected to improve the creep strength of this steel.

Keywords: creep resistant steels; carbonitrides precipitation; martensite; tempering; thermomechanical treatment; ferritic/martensitic steel; MX nanoprecipitates

1. Introduction

The 9-12Cr Ferritic/martensitic (FM) steels are widely used in power generating and petrochemical plants for operating temperatures up to 620 °C because of their excellent combination of creep strength, thermal properties, and oxidation resistance as well as acceptable room temperature mechanical properties and also good weldability [1]. In the designs of advanced power plant components for future fission and fusion power plants, which will operate at temperatures up to 650 °C, the primary emphasis was placed on the development of new steel grades with superior long-term creep and thermal fatigue properties [2,3]. Several promising ferritic steels have been developed by the addition of very expensive alloying elements such as W and Co [4–7]. However, there are economic and technical advantages for using the same basic composition of ASTM T/P 91 (here after named G91) and considerably raising creep-strength of this material by the development of thermomechanical treatments (TMT) [8–11]. The creep resistance of a G91 steel results from the combination of several types of barriers to dislocation motion: a high density of martensitic lath boundaries, solid solution strengthening with elements with much larger atomic size than iron, and fine dispersion of second phase particles [12,13]. Although the overall creep strain rate is the result of the combined effects of these mechanisms, it has been reported in previous works that the creep strength is mainly increased by the precipitation of fine MX carbonitrides in the matrix [14,15], which are very effective barriers to pin dislocations, and very stable during long-term aging [16–18].

The goal of this paper is to develop a processing route to improve the creep strength of a conventional G91 steel. Prior to the optimization of TMT, thermodynamic calculations using ThermoCalc® will be carried out to foresee the most promising microstructures. In order to maximize the number of nanoscale MX precipitates, an ausforming procedure will be applied to increase the number of nucleation sites for precipitation inside the martensite lath relative to standard heat treatments. Previous works have investigated the effect of applying a TMT instead of a conventional heat treatment demonstrating the improvement in creep strength achieved with a finer dispersion of MX [8,9,11]. This work tries to deepen our understanding of the effect of the ausforming temperature on the tempered martensitic microstructure in order to quantify the importance of ausforming in the TMT to optimize the creep strength. The conclusions achieved in this work will allow us to clarify what processing parameters affect the heterogeneous formation of thermally stable precipitates.

2. Materials and Methods

A commercial G91 FM steel supplied by CIEMAT (Madrid, Spain) in the form of a plate was used in this work. The nominal chemical composition of this steel is given in Table 1. This material was received after a heat treatment consisting of normalization at 1040 °C for 30 min followed by tempering at 730 °C for 1 h, both with air cooling to room temperature.

Table 1. Nominal composition in wt % of Grade 91 ferritic/martensitic steel.

Elements	C	Si	Mn	Cr	Mo	V	Nb	N	Fe
Wt %	0.1	0.4	0.4	9.0	1.0	0.2	0.07	0.038	balance

Thermodynamic calculations, used as guidelines to select the temperatures for the Thermomechanical Treatment (TMT), were performed with the Thermo-Calc® (Solna, Sweden) software based on the CALPHAD (Computer Coupling of Phase Diagrams and Thermo-chemistry) using the database TCFE8. The new TMT approach proposed in this work is shown schematically in Figure 1. Cylindrical samples with 10 mm length × 5 mm diameter were given 20% deformation at $0.1\ s^{-1}$ in a Bähr DIL 805A/D plastodilatometer (TA Instruments, New Castle, DE, USA). The samples were heated at 5 °C/s and cooled at 50 °C/s.

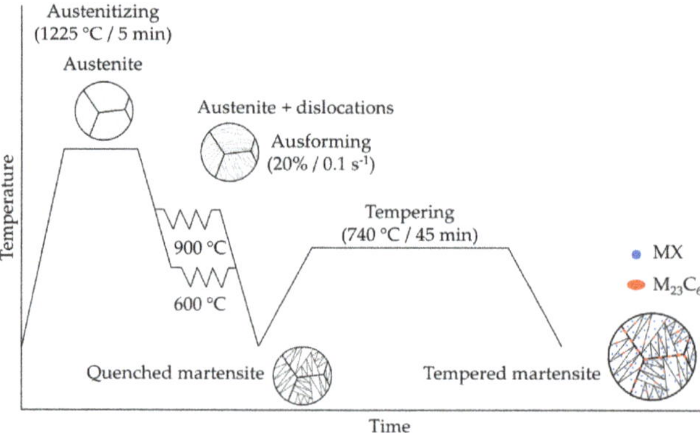

Figure 1. Thermomechanical Treatment (TMT) scheme.

The microstructure of as-received material and material after TMT simulation treatments was analyzed using optical microscopy, field emission gun scanning electron microscopy (FEG-SEM), orientation maps obtained by electron backscatter diffraction (EBSD), transmission electron microcopy (TEM), and X-ray diffraction (XRD). Standard grinding and polishing procedures were used for sample preparation, which included a final polish with 1 micron diamond paste. Polished specimens were etched with a solution of 5 mL hydrochloric acid, 1 g of picric acid and 100 mL of ethyl alcohol (Vilella's reagent) to develop the microstructural features.

EBSD measurements were performed with a JEOL JSM 6500 FEG-SEM (JEOL Ltd., Tokyo, Japan) operating at 20 kV equipped with a fully automatic EBSD attachment from Oxford Instruments HKL (Abingdon, UK). Residual damage in EBSD samples from diamond polishing was removed through an additional polishing stage with colloidal (40 nm). EBSD mapping has been carried out on an area of about $900 \times 400\ \mu m^2$ at step sizes of 0.4 μm. The HKL Channel 5 software (Oxford Instruments, Abingdon, UK) has been used for data processing in order to obtain a graphical representation of the microstructure by marking the sample points of the grains in a map using colors specific to the lattice orientation.

Observation of precipitates present in the microstructure was performed on thin foils by TEM in a JEOL JEM 2100 and a JEOL JEM 3000F (JEOL Ltd., Tokyo, Japan). For this goal, disks 3 mm in diameter and 100 μm in thickness were cut. The thickness of these slices were further mechanically thinned and, afterwards, twin-jet electropolishing was performed at 25 °C and 40 V using an electrolyte compose of 95% acetic acid and 5% percloric acid.

XRD studies were carried out with Co-K$_\alpha$ radiation in a Bruker AXS D8 diffractometer equipped with Goebel mirror optics and a LynxEye Linear Position Sensitive Detector for ultra-fast XRD measurements. For the Rietveld refinement of the diffractograms, the version 4.2 program TOPAS (4.2, Bruker AXS, Billerica, MA, USA) has been used and the crystallographic information of the phases used was obtained from Pearson's Crystal Structure Database for Inorganic Compounds, Release 2015/2016. Materials Park: ASM International, 2015. Line broadening due to the crystallite size and lattice strain was analyzed using the double-Voigt approach. It was concluded from this analysis that line broadening is mainly due to the microstrain produced by dislocations and their associated stress fields.

Finally, Vickers hardness measurements were performed with a load of 5 kg on the longitudinal section of warm/hot compressed material before and after tempering.

3. Results and Discussion

3.1. Thermodynamic Calculations

The equilibrium phase mole fraction of the coexisting phases in the microstructure of the G91 FM steel is shown in Figure 2 as a function of temperature. According to this figure, $M_{23}C_6$ carbides will dissolve during the conventional austenitizing treatment at 1040 °C, remaining undissolved about 1.59×10^{-3} mole fraction of MX precipitates. In order to improve the creep strength during the service life, it is necessary to maximize the amount of MX carbides finely dispersed that precipitate during the TMT. This point emphasizes the importance of increasing the austenitizing temperature to guarantee the dissolution of most MX precipitates present in the as-received microstructure. However, this treatment must be performed in the fully austenite region to avoid the presence of δ-ferrite, which is generally regarded as detrimental for creep properties [19]. Since δ-ferrite forms above 1255 °C, according to Figure 2, an austenitizing temperature of 1225 °C was selected. At this temperature, 5 min treatment was enough to get a complete homogenization of the microstructure due to the small size of the sample used. The effect of increasing the austenitizing temperature on the precipitation reactions during tempering was evaluated using Thermo-Calc®. The volume fraction of MX carbides (with M = Nb, V and X = C, N) present in the austenite decreases from 1.32×10^{-3} to 1.01×10^{-4} when the temperature is increased from 1040 to 1225 °C. Therefore, more Nb/V/C/N will be available in solid solution in the austenite at 1225 °C, compared to 1040 °C, to precipitate during the subsequent tempering step. Thus, as some MX precipitates will still remain undissolved after austenitizing, the composition (of the martensite) prior to the calculation at the tempering step has to be corrected. These calculations show an increase in the volume fraction of MX precipitates during tempering at 740 °C from 1.42×10^{-3} to 2.67×10^{-3} when the austenitizing temperature is increased from 1040 to 1225 °C (two times greater). It should be kept in mind when comparing these calculations to the experimental results that the authors assumed that quasi-equilibrium has been reached during the austenitizing and tempering steps investigated in this work.

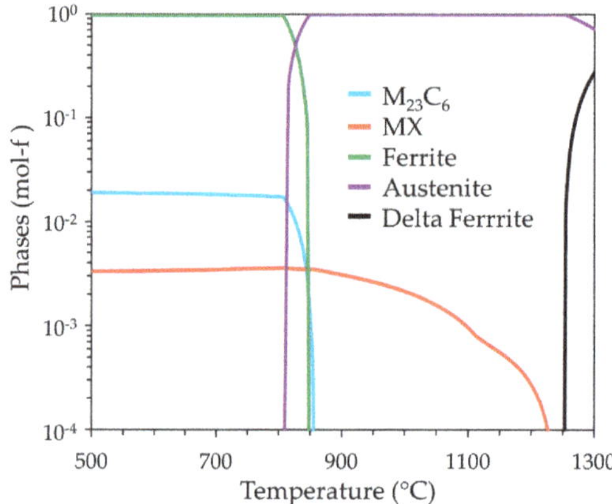

Figure 2. Temperature evolution of phase mole fraction in G91 calculated by ThermoCalc.

3.2. Microstructural Characterization

As shown in the micrograph of Figure 3a, as-received G91 steel presents a tempered martensite microstructure, with grain and lath boundaries decorated by tiny bright spots corresponding to M$_{23}$C$_6$ particles precipitated during tempering. Quantitative metallography studies revealed an average lath size ranging from 0.25 to 0.5 µm [20]. Examination of this microstructure at higher magnifications with a TEM reveals the presence of MX precipitates, distributed in the matrix within the lath, in addition to M$_{23}$C$_6$, distributed only on boundaries. As shown in Figure 3b,c, the size of these precipitates ranges from 100 to 200 nm for M$_{23}$C$_6$ carbides and from 20 to 50 nm for MX precipitates.

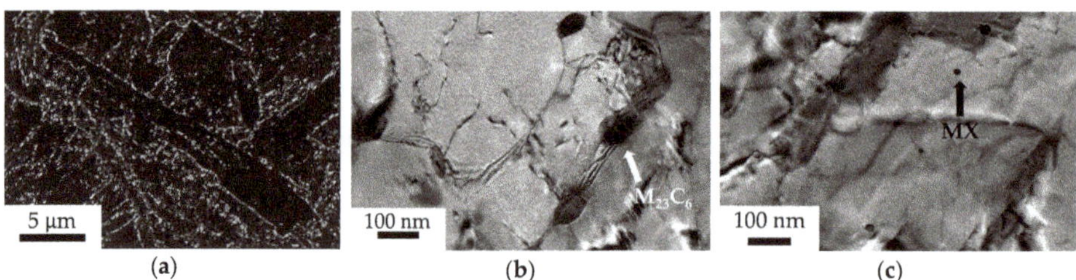

Figure 3. Microstructure of G91 after conventional heat treatment showing M$_{23}$C$_6$ precipitates (**a**) on grain boundaries (Scanning electron microscopy (SEM) micrograph) and (**b**) on lath boundaries (Transmission electron microcopy (TEM) micrograph) and (**c**) MX within laths (TEM micrograph).

SEM examination showed that, during the austenitizing treatment at 1225 °C, all carbides were dissolved except for a negligible volume fraction of Nb rich MX precipitates (Figure 4).

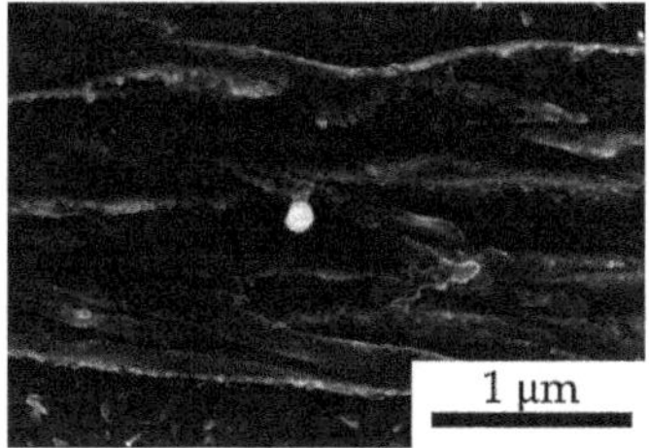

Figure 4. SEM micrographs showing Nb rich MX precipitate after austenitizing treatment at 1225 °C.

The XRD pattern obtained in a sample quenched to room temperature after the austenitization showed only the diffraction peaks of ferrite (Figure 5). No retained austenite was detected in the microstructure by this technique after the austenitization heat treatment.

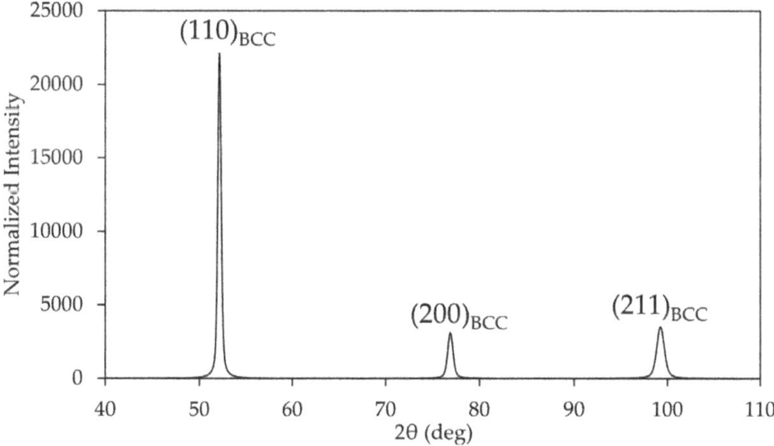

Figure 5. X-ray diffraction pattern after austenitizing at 1225 °C. BCC stands for body-centered cubic ferrite.

Comparing the average prior austenite grain size present in as-received material (43 μm) and after austenitizing at 1225 °C (259 μm), it is observed that a considerable grain growth occurs (Figure 6). No δ-ferrite has been formed after austenitizing at 1225 °C. The inhibition of the austenite grain growth during austenitizing is related to the presence of fine second phase particles, dispersed in the microstructure, that pin the grain boundaries and restrain their movement. The explosive growth observed indicates that around 1225 °C most of the MX and $M_{23}C_6$ precipitates are dissolved. In general, a coarse-grained microstructure presents better creep properties. High-temperature creep is controlled by diffusion, and thus creep strength increases with a decreasing the amount of regions with a high diffusion rate like grain boundaries.

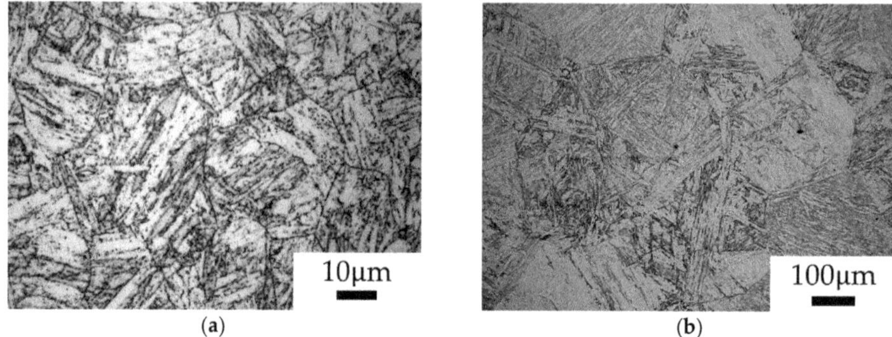

Figure 6. Optical micrographs showing the prior austenite grain microstructure in G91: (**a**) after conventional heat treatment and (**b**) after austenitizing at 1225 °C.

To maximize the number of nanoscale MX precipitates, a thermomechanical procedure was used to increase the dislocation density in the steel, increasing the number of nucleation sites for precipitation inside the martensite laths during the subsequent tempering stage at 740 °C. After austenitizing at 1225 °C, samples were cooled to either 900 or 600 °C and hot-worked in a controlled manner. Hot deformation at 900 °C results in dynamic recrystallization (DRX) as can be seen in Figure 7, leading to grain refinement, from 259 μm for austenitized at 1225 °C material to 141 μm. On the other hand, as the transformation of austenite to ferrite at 600 °C would take place after a long time, metastable austenite can be plastically deformed at this temperature without inducing this phase transformation. In this case, the low ausforming temperature and high strain rate would suggest that neither recrystallization nor recovery of the microstructure is taking place and, thus, a more refined martensitic structure in the quenched samples is expected.

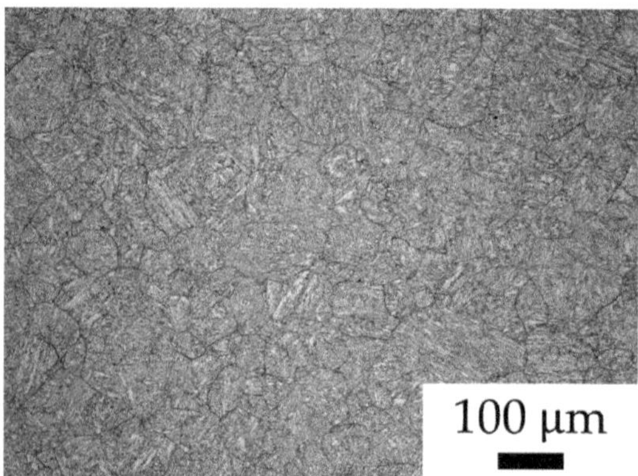

Figure 7. Optical micrographs showing the prior austenite grain microstructure in G91 after ausfroming at 900 °C.

EBSD investigations revealed a significant difference in the substructure of the samples ausformed at 600 and 900 °C, as shown in Figure 8a,b, respectively. While samples deformed at 900 °C showed the typical lath martensite structure of blocks (Figure 8b), this lath-like morphology after deformation at 600 °C is blurred (Figure 8a). This is attributed to the higher number density of dislocations introduced during the deformation of austenite at 600 °C, which hindered typical block growth during the martensitic transformation. Table 2 collects the block width measured on three IPF (Inverse pole figure) maps obtained by EBSD for each sample by the linear intercept method [21]. Boundaries with a misorientation larger than 10 were considered as block boundaries in the measurements. Our observations reveal coarser blocks at the lower ausforming temperature. This is consistent with the variant selection concept reported by other authors [22,23], where the strengthening of austenite enhances self-accommodation of the transformation strain by formation of preferential variants. Compared to the G91 in as-received condition (Figure 8c), whose values were also measured (Table 2), a higher block width is obtained for the samples after TMT due to the higher austenitization temperature used in the TMT.

Table 2 shows the lath width measured on eight TEM micrographs for each microstructures by the linear intercept method. These results demonstrate that the lath width in the G91 in as-received conditions can be refined by TMT, and that the decrease of the ausforming temperature leads to a decrease of the martensite lath width. This fact is easily explainable because the lath width depends on the strength of austenite and, thus, lower ausforming temperatures enable greater austenite strengthening [24].

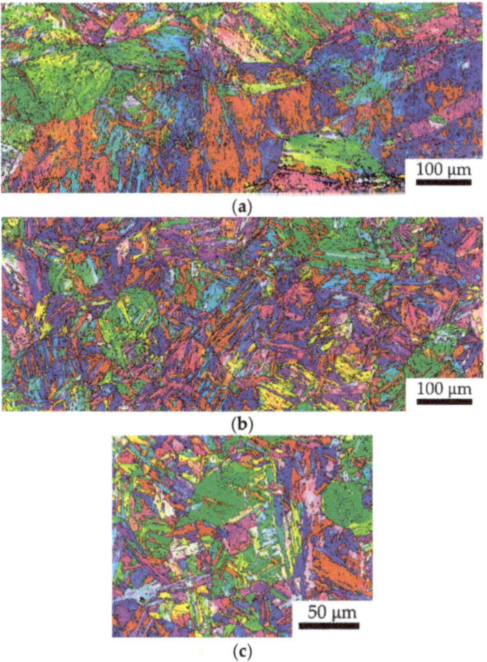

Figure 8. IPF (Inverse pole figure) maps after TMT for the sample (**a**) ausformed at 600 °C, (**b**) ausformed at 900 °C, (**c**) G91 in as-received condition

Table 2. Block and lath width for the samples after TMT and G91 in as-received condition.

Sample	Block Width (μm)	Lath Width (nm)
G91 as-received	2.71 ± 0.23	356 ± 35
Def. at 900 °C	3.23 ± 0.26	285 ± 26
Def. at 600 °C	3.91 ± 0.36	212 ± 59

Table 3 shows that hardness of as-quenched sample is increased by the ausforming treatment. The higher hardness value obtained for the sample ausformed at 600 °C could be justified on the basis of having a higher dislocation density and a finer martensite lath-structure, which is related to the lack of dynamic recrystallization during ausforming at 600 °C.

Table 3. Hardness values after austenitization and ausforming at different temperatures.

No Def.	Def. at 900 °C	Def. at 600 °C
426 ± 3	434 ± 5	465 ± 10

As shown in Table 4, ausforming causes a decrease in the martensite start (Ms) and finish (Mf) temperatures, resulting from the mechanical stabilization of austenite produced when the strength of austenite is enough to avoid the motion of glissile interfaces [25]. Therefore, the higher dislocation density obtained for the sample ausformed at 600 °C is translated into having the lowest transformation temperature.

Table 4. Martensitic transformation temperatures after austenitization and ausforming at different temperatures and quenching (50 °C/s) to room temperature.

Sample	Ms (°C)	Mf (°C)	ΔM (°C)
Austenitized	385	220	165
Def. at 900 °C	374	195	179
Def. at 600 °C	338	145	193

Tempering is the key step of the TMT to achieve the optimum precipitate size, and the recovery degree of the martensitic matrix that guarantees the optimal combination of toughness and creep strength properties. As commercially used for this steel grade, a tempering temperature of 740 °C and a holding time of 45 min were selected in this work. To understand the effect of the TMT applied, the number density and average size of precipitates were calculated analyzing at least 100 precipitates for MX from TEM micrographs and 300 $M_{23}C_6$ from SEM micrographs. As shown in Table 5, the ausforming treatment does not produce a significant change in size and distribution of the $M_{23}C_6$. The precipitation processes of these carbides take place on prior austenite grain, block, and lath boundaries as soon as the tempering begins, and it is completed in a short time [26] (Figure 9a,b,d,e). The lath refinement produced during ausforming can increase the number of potential nucleation sites for these precipitates on lath boundaries. By contrast, greater block width and prior austenite grain is obtained after TMT. Since $M_{23}C_6$ precipitates nucleate first on prior austenite grain and block boundaries due to its higher energy, the remaining carbon content will precipitate as $M_{23}C_6$ on lath boundaries obtaining finer $M_{23}C_6$ distribution in TMT samples. This difference in size between the $M_{23}C_6$ formed on lath boundaries, and those nucleated on block and prior austenite grain boundaries, may produce a faster Ostwald ripening coarsening resulting in an average $M_{23}C_6$ size and number density similar to the G91 in as-received condition. Besides, the number density of MX precipitates in the TMT samples increases in three orders of magnitude, having a size about four times smaller than those in the G91 in as-received condition, which is in good agreement with previous works where similar thermomechanical treatment were applied in G91 [8,14] (Figure 9c,f). The higher austenitization temperature and ausforming processing greatly increased the dislocation density in austenite as others author have demonstrated in previous works [27], resulting in a higher number of nucleation sites for precipitation of MX precipitates inside the martensite laths. Thus, the higher dislocation density introduced by deformation at 600 °C is responsible for a smaller size of these precipitates and the slightly higher number density. Partial recrystallization during ausforming at 900 °C causes a significant decrease in dislocation density in recrystallized grains as well as an inhomogeneous distribution of MX precipitate.

Table 5. Number density and size of precipitates after TMT and G91 in as-received condition.

Sample	Precipitate	Diameter (nm)	Number Density (m^{-3})
G91 as-received	$M_{23}C_6$	141 ± 4	6.20^{19}
	MX	25 ± 0.6	8.1410^{19}
Def. at 900 °C	$M_{23}C_6$	125 ± 3	8.5010^{19}
	MX	7.4 ± 0.3	6.410^{22}
Def. at 600 °C	$M_{23}C_6$	136 ± 5	3.7810^{19}
	MX	5.59 ± 0.4	9.3910^{22}

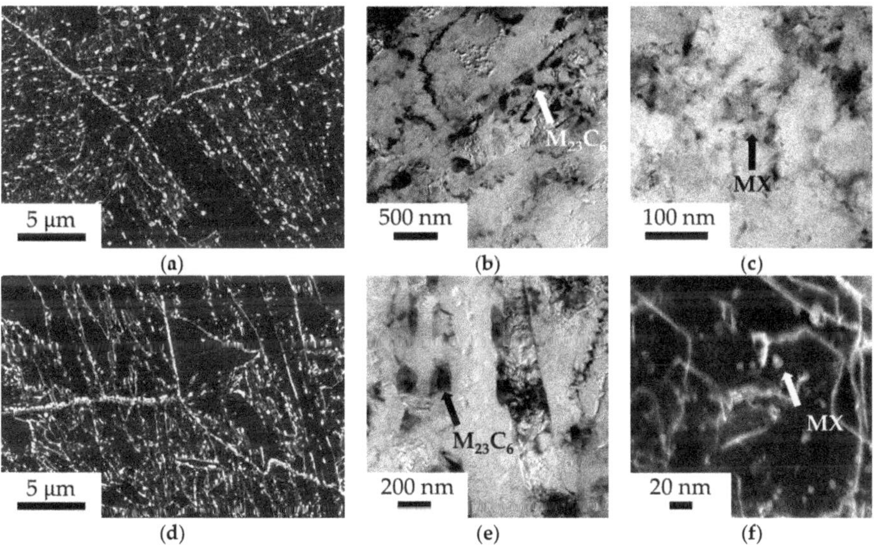

Figure 9. (**a**) SEM micrograph showing $M_{23}C_6$ on grain boundaries and TEM micrographs showing, (**b**) $M_{23}C_6$ on lath boundaries, and (**c**) MX within laths for the sample after TMT (ausformed at 600 °C); and (**d**) SEM micrograph and TEM micrographs showing (**e**) $M_{23}C_6$ on lath boundaries and (**f**) MX within laths for the sample after TMT (ausformed at 900 °C).

Hardness measurements were performed in the G91 in as-received condition and after TMTs (Table 6). The samples after TMTs show hardness values up to 50 HV5 higher than the ones for G91 in as-received condition, being the values for the TMT samples very similar. These results together with the microstructural characterization carried out demonstrate the improvement in strength after TMT, which is attributable to the lath refinement along with the increase in number density and reduction in size of MX precipitates.

Table 6. Hardness values for the sample after TMT and G91 in as-received condition.

G91 as-Received	Def. at 900 °C	Def. at 600 °C
260 ± 2	310 ± 9	$307 + 8$

4. Conclusions

The microstructural analyses carried out allow us to conclude that applying a TMT instead of conventional heat treatment on G91 ferritic/martensitic steel will promote:

1. Microstructures containing three orders of magnitude higher number density and four times smaller size of MX precipitates.
2. Microstructures contain $M_{23}C_6$ precipitates with very similar values of size and number density.
3. An increase in hardness of 50 HV5, which is attributable to the increase in the number density and the reduction in the size of MX nanoprecipitates and the lath width refinement.

The main conclusions reached on the effect of ausforming on the martensitic microstructure are summarized below:

4. The martensitic microstructure reduces its lath width due to the strengthening of austenite produced by ausforming. However, ausforming at 600 °C results in a higher block width as compared to ausfoming at 900 °C because of the variant selection that takes place during the martensitic transformation in the sample ausformed at 600 °C.
5. The ausforming temperature does not affect $M_{23}C_6$ distribution and size, but ausforming at 600 °C leads to a slightly smaller size of MX.

Acknowledgments: The authors acknowledge financial support to Spanish Ministerio de Economia y Competitividad (MINECO) in the form of a Coordinate Project (MAT2016-80875-C3-1-R). The work presented here is done within the Joint Program on Nuclear Materials of the European Energy Research Alliance Pilot Project CREMAR. The authors also would like to acknowledge financial support to Comunidad de Madrid through DIMMAT-CM_S2013/MIT-2775 project. The authors are grateful to Javier Vara, Alberto Delgado, and Alberto López for the experimental support. Javier Vivas acknowledges financial support in the form of a FPI Grant BES-2014-069863.

Author Contributions: All authors were involved in discussing the results and in finalizing the manuscript.

Conflicts of Interest: The authors declare no conflict of interest.

References

1. Mayer, K.H.; Masuyama, F. The development of creep-resistant steels. In *Creep-Resistant Steels*; Woodhead Publishing: Cambridge, UK, 2008; pp. 15–77.
2. Klueh, R.L.; Gelles, D.S.; Jitsukawa, S.; Kimura, A.; Odette, G.R.; van der Schaaf, B.; Victoria, M. Ferritic/martensitic steels—Overview of recent results. *J. Nucl. Mater.* **2002**, *307*, 455–465. [CrossRef]
3. Klueh, R.L.; Ehrlich, K.; Abe, F. Ferritic/martensitic steels: Promises and problems. *J. Nucl. Mater.* **1992**, *191*, 116–124.
4. Helis, L.; Toda, Y.; Hara, T.; Miyazaki, H.; Abe, F. Effect of cobalt on the microstructure of tempered martensitic 9Cr steel for ultra-supercritical power plants. *Mater. Sci. Eng. A* **2009**, *510*, 88–94. [CrossRef]
5. Kipelova, A.; Kaibyshev, R.; Belyakov, A.; Molodov, D. Microstructure evolution in a 3% co modified P911 heat resistant steel under tempering and creep conditions. *Mater. Sci. Eng. A* **2011**, *528*, 1280–1286. [CrossRef]
6. Knežević, V.; Balun, J.; Sauthoff, G.; Inden, G.; Schneider, A. Design of martensitic/ferritic heat-resistant steels for application at 650 °C with supporting thermodynamic modelling. *Mater. Sci. Eng. A* **2008**, *477*, 334–343. [CrossRef]
7. Rojas, D.; Garcia, J.; Prat, O.; Sauthoff, G.; Kaysser-Pyzalla, A.R. 9%Cr heat resistant steels: Alloy design, microstructure evolution and creep response at 650 °C. *Mater. Sci. Eng. A* **2011**, *528*, 5164–5176. [CrossRef]
8. Klueh, R.L.; Hashimoto, N.; Maziasz, P.J. Development of new nano-particle-strengthened martensitic steels. *Scr. Mater.* **2005**, *53*, 275–280. [CrossRef]
9. Hollner, S.; Fournier, B.; Le Pendu, J.; Cozzika, T.; Tournié, I.; Brachet, J.C.; Pineau, A. High-temperature mechanical properties improvement on modified 9Cr–1mo martensitic steel through thermomechanical treatments. *J. Nucl. Mater.* **2010**, *405*, 101–108. [CrossRef]
10. Tan, L.; Busby, J.T.; Maziasz, P.J.; Yamamoto, Y. Effect of thermomechanical treatment on 9Cr ferritic–martensitic steels. *J. Nucl. Mater.* **2013**, *441*, 713–717. [CrossRef]
11. Hollner, S.; Piozin, E.; Mayr, P.; Caës, C.; Tournié, I.; Pineau, A.; Fournier, B. Characterization of a boron alloyed 9Cr3W3CoVNbBN steel and further improvement of its high-temperature mechanical properties by thermomechanical treatments. *J. Nucl. Mater.* **2013**, *441*, 15–23. [CrossRef]

12. Abe, F. Precipitate design for creep strengthening of 9% Cr tempered martensitic steel for ultra-supercritical power plants. *Sci. Technol. Adv. Mater.* **2008**, *9*, 013002. [CrossRef] [PubMed]
13. Maruyama, K.; Sawada, K.; Koike, J.I. Strengthening mechanisms of creep resistant tempered martensitic steel. *ISIJ Int.* **2001**, *41*, 641–653. [CrossRef]
14. Klueh, R.L.; Hashimoto, N.; Maziasz, P.J. New nano-particle-strengthened ferritic/martensitic steels by conventional thermo-mechanical treatment. *J. Nucl. Mater.* **2007**, *367*, 48–53. [CrossRef]
15. Li, S.; Eliniyaz, Z.; Sun, F.; Shen, Y.; Zhang, L.; Shan, A. Effect of thermo-mechanical treatment on microstructure and mechanical properties of p92 heat resistant steel. *Mater. Sci. Eng. A* **2013**, *559*, 882–888. [CrossRef]
16. Abe, F. Coarsening behavior of lath and its effect on creep rates in tempered martensitic 9Cr-W steels. *Mater. Sci. Eng. A* **2004**, *387*, 565–569. [CrossRef]
17. Tamura, M.; Sakasegawa, H.; Kohyama, A.; Esaka, H.; Shinozuka, K. Effect of mx type particles on creep strength of ferritic steel. *J. Nucl. Mater.* **2003**, *321*, 288–293. [CrossRef]
18. Tan, L.; Byun, T.S.; Katoh, Y.; Snead, L.L. Stability of mx-type strengthening nanoprecipitates in ferritic steels under thermal aging, stress and ion irradiation. *Acta Mater.* **2014**, *71*, 11–19. [CrossRef]
19. Kimura, K.; Sawada, K.; Kushima, H.; Toda, Y. Influence of chemical composition and heat treatment on long-term creep strength of grade 91 steel. *Procedia Eng.* **2013**, *55*, 2–9. [CrossRef]
20. Klueh, R.L. Elevated temperature ferritic and martensitic steels and their application to future nuclear reactors. *Int. Mater. Rev.* **2005**, *50*, 287–310. [CrossRef]
21. Caballero, F.G.; Roelofs, H.; Hasler, S.; Capdevila, C.; Chao, J.; Cornide, J.; Garcia-Mateo, C. Influence of bainite morphology on impact toughness of continuously cooled cementite free bainitic steels. *Mater. Sci. Technol.* **2012**, *28*, 95–102. [CrossRef]
22. Shi, Z.; liu, K.; Wang, M.; Shi, J.; Dong, H.; Pu, I.; Chi, B.; Zhang, Y; Jian, L. Effect of tensile deformation of austenite on the morphology and strength of lath martensite. *Met. Mater. Int.* **2012**, *18*, 317–320. [CrossRef]
23. Miyamoto, G.; Iwata, N.; Takayama, N.; Furuhara, T. Variant selection of lath martensite and bainite transformation in low carbon steel by ausforming. *J. Alloys Compd.* **2013**, *577*, S528–S532. [CrossRef]
24. Zhang, M.; Wang, Y.H.; Zheng, C.L.; Zhang, F.C.; Wang, T.S. Austenite deformation behavior and the effect of ausforming process on martensite starting temperature and ausformed martensite microstructure in medium-carbon si–al-rich alloy steel. *Mater. Sci. Eng. A* **2014**, *596*, 9–14. [CrossRef]
25. Chatterjee, S.; Wang, H.S.; Yang, J.R.; Bhadeshia, H.K.D.H. Mechanical stabilisation of austenite. *Mater. Sci. Technol.* **2006**, *22*, 641–644. [CrossRef]
26. Prat, O.; García, J.; Rojas, D.; Sanhueza, J.P.; Camurri, C. Study of nucleation, growth and coarsening of precipitates in a novel 9%Cr heat resistant steel: Experimental and modeling. *Mater. Chem. Phys.* **2014**, *143*, 754–764. [CrossRef]
27. Seo, S.W.; Jung, G.S.; Lee, J.S.; Bae, C.M.; Bhadeshia, H.K.D.H.; Suh, D.W. Ausforming of medium carbon steel. *Mater. Sci. Technol.* **2015**, *31*, 436–442. [CrossRef]

Technical Note

Tensile Ductility of Nanostructured Bainitic Steels: Influence of Retained Austenite Stability

Thomas Sourmail [1,*], Carlos Garcia-Mateo [2], Francisca G. Caballero [2], Lucia Morales-Rivas [3], Rosalia Rementeria [2] and Matthias Kuntz [4]

[1] Ascometal CREAS, Avenue de France, 57300 Hagondange, France
[2] Department of Physical Metallurgy, National Center for Metallurgical Research (CENIM-CSIC), Avda. Gregorio del Amo, 8, E-28040 Madrid, Spain; cgm@cenim.csic.es (C.G.-M.); fgc@cenim.csic.es (F.G.C.); rosalia.rementeria@cenim.csic.es (R.R.)
[3] University of Kaiserslautern, Materials Testing, Gottlieb-Daimler-Str., 67663 Kaiserslautern, Germany; rivas@mv.uni-kl.de
[4] Robert Bosch GmbH, Materials- and Process Engineering Metals, Renningen, 70465 Stuttgart, Germany; matthias.kuntz2@de.bosch.com
*   Correspondence: thomas.sourmail@ascometal.com; Tel.: +33-3-87-70-73-16

Academic Editor: Hugo F. Lopez
Received: 19 December 2016; Accepted: 16 January 2017; Published: 23 January 2017

Abstract: High silicon (>1.5%) steels with different compositions were isothermally transformed to bainite at 220 and 250 °C to produce what is often referred to as nanostructured bainite. Interrupted tensile tests were carried out and the retained austenite was measured as a function of strain. Results were correlated with tensile ductility. The role of retained austenite stability is remarkably underlined as strongly affecting the propensity to brittle failure, but also the tensile ductility. A simple quantitative relationship is proposed that clearly delimitates the different behaviours (brittle/ductile) and correlates well with the measured ductility. Conclusions are proposed as to the role of retained austenite fraction and the existence of a threshold value associated with tensile rupture.

Keywords: high carbon steels; nanobainite; low temperature bainite; tensile ductility; retained austenite stability; transformation induced plasticity (TRIP)

---

1. Introduction

Bainitic microstructures formed at low temperatures (350 °C or less) have received a considerable amount of attention in the recent years [1–14]. These microstructures are obtained in relatively high carbon steels (0.6–1.2 wt %, although the concept can be extended to lower carbon contents) through isothermal transformation over durations ranging from 10 to over 100 h [5,11,15]. They consist of ultrafine bainitic laths (typical width under 50 nm) surrounded by retained austenite [3,11]. Interestingly, the initial mechanical properties [2] were at best on par with those of quenched and tempered high strength spring steels [16], but were later improved to reach an unprecedented 21% elongation for over 2.1 GPa in tensile strength [10,11].

From a microstructural point of view, the yield and tensile strength of these materials have been shown to be reasonably well correlated to the parameter V_β / t_β where V_β is the volume fraction of bainitic ferrite (the rest normally being retained austenite) and t_β the average lath thickness [10,11,17]. However, tensile ductility has recently been shown to exhibit largely different values for microstructures exhibiting reasonably similar retained austenite fraction and bainitic ferrite lath thickness, as shown in Table 1, after [12].

Table 1. Tensile elongation for an identical material (1C-2.5Si wt %) transformed at 220 or 250 °C, after [12].

Reference in [12]	UTS, MPa	TE, %	$\gamma_{res} = 1 - V_\beta$, %	t_β, nm
1CSi-220	~2070	7	34	28
1CSi-250	~2200	21	36	28

UTS is Ultimate tensile strength; TE is Total elongation; γ_{res} is the volume fraction of retained austenite; V_β is the volume fraction of bainitic ferrite; t_β is the average thickness of the bainitic ferrite laths.

Earlier work on the factors controlling the ductility of nanostructured bainite has insisted on the role of retained austenite fraction [18,19] and suggested the existence of an optimum value to achieve the maximum ductility. More recent work, based on measurements of retained austenite content before tensile tests and calculated evolutions, proposed that the stability of retained austenite would influence the material ductility, and indicated that there could be an optimum stability [20]. Data from this same publication were later re-interpreted to propose the existence of a percolation mechanism, whereby ductility was imparted by a percolating network of retained austenite in the matrix of bainitic ferrite, and fracture occurred at an approximately constant volume fraction of 10% retained austenite [17]. Also recently, a model for stress-assisted martensite formation was proposed by the present authors [21] and was found to provide a reasonable agreement with the experiment, though this approach does not provide indication as to the causes of early tensile failures.

The present work is concerned with further investigating this hypothesis through the use of interrupted tensile tests. This allows actual measurement of the retained austenite content as a function rather than estimated values to be used.

2. Materials and Methods

2.1. Materials

Three materials were used for the present investigation, with compositions as indicated in Table 2. References indicate both the carbon and silicon content. Both 06C-1.5Si and 1C-2.5Si were produced industrially as ingots, then hot-rolled to 120 mm (0.6C-1.5Si) or 35 mm (1C-2.5Si) bars. The 1C-1.5Si steel was manufactured using a vacuum induction furnace to obtain an approximately 35 kg ingot. After cooling to room temperature, the ingot was re-heated to 1150 °C and forged to a 40 mm bar. Prior to machining, all steels were annealed for 2 h at 700 °C. Chemical composition was determined on the hot-rolled or forged bars using optical emission spectrometry and combustion analysis (LECO).

Table 2. Chemical composition (wt %) for the three steels used in the present investigation, as determined using optical emission spectrometry and combustion (LECO) analysis.

Reference	C	Si	Mn	Cr	Mo	V
0.6C-1.5Si	0.67	1.67	1.32	1.73	0.15	0.12
1C-1.5Si	1.05	1.60	0.74	1.05	0.07	-
1C-2.5Si	0.99	2.47	0.74	0.97	0.03	-

2.2. Heat-Treatment, Tensile Testing and Retained Austenite Measurements

Tensile specimens, 6 or 8 mm in diameter, were manufactured from the hot-rolled bars, using material taken at mid-radius of the latter. They were initially machined with 0.3–0.5 mm additional thickness then heat-treated. The heat-treatments consisted of austenitising in a first salt bath or in a conventional furnace, followed by rapid cooling in a salt bath, to the isothermal transformation temperature. Both austenitising and isothermal transformation parameters varied. Austenitising was carried out for 1 h at temperatures between 860 and 1050 °C, while the temperature for isothermal transformation varied between 220 and 250 °C. The duration for isothermal holding was determined from measurements in a Baehr dilatometer and varied depending on the material and austenitising. For convenience, relevant heat-treatment parameters are included in the specimen reference. As an example, 1C-2.5Si-1050-250 (16 h) refers to material 1C-2.5Si austenitised at 1050 °C for one hour and isothermally transformed at 250 °C for 16 h; austenitising duration is not indicated as it was kept constant (1 h). Following heat-treatment, specimens were hard-machined to their final dimensions.

Conventional tensile tests were carried out using three to five specimens. Once yield strength, universal tensile strength and tensile elongation were known, interrupted tensile tests were carried out at selected values of plastic strain in the uniform elongation domain (so as to ensure absence of necking and non-uniform strain distribution in the specimens). Retained austenite measurements were carried out on both the tensile specimens' grip (reference value) and on transverse sections from the gauge length (value after destabilization by plastic strain).

For these experiments, samples were machined, ground and polished with 1 μm diamond paste, and then subjected to several cycles of etching and polishing to obtain an undeformed surface; finally, the samples were polished with colloidal silica. X-ray diffraction measurements were performed by means of a Bruker AXS D8 diffractometer equipped with a Co X-ray tube and Goebel mirror optics to obtain a parallel and monochromatic X-ray beam. Operational parameters and the procedure for obtaining the austenite content and composition are described elsewhere [22,23].

3. Results

Tensile tests exhibited two different behaviors which are illustrated in Figure 1. For specimens breaking in a brittle manner, both ultimate tensile strength (UTS) and elongation varied significantly (up to 800 MPa for the maximum stress and 4%–5% for the maximum elongation) in the three to five tests carried out on each identical condition, the maximum values were taken. For specimens breaking after necking, the reproducibility was typically within ±15 MPa for the UTS and ±0.7% for the elongation.

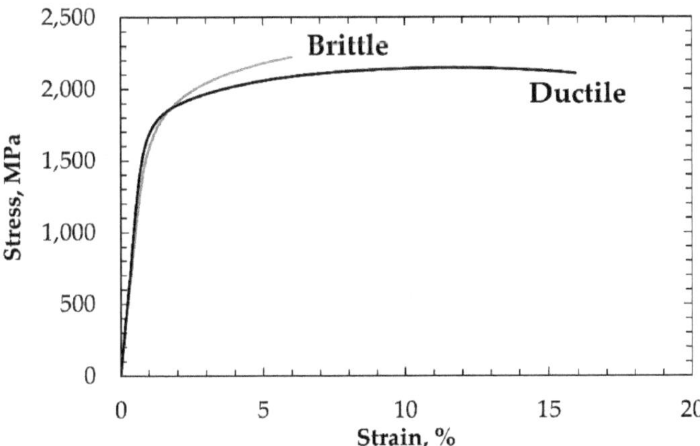

Figure 1. Example of the two different behaviours (brittle and ductile) identified on engineering stress–strain curves. Brittle behaviour may lead to higher maximum stress though the reproducibility is poor.

Tensile data for all investigated conditions are summarized in Table 3, together with retained austenite content as measured in the grip. As shown, UTS varied between 2.0 and 2.2 GPa, with elongation as high as 17%. As already reported [10,11], such results are exceptional in the combination of strength and ductility that is achieved. It is also clear that retained austenite content alone does not correlate directly with tensile elongation, as elongations around 5% and above 10% can be found for both ~20% or ~40% retained austenite contents.

Table 3. Initial retained austenite content and tensile properties for all conditions investigated. The notation for the reference is explained in the text. * indicate brittle behaviour with no true UTS value and variable maximum elongation (the highest value of all tests is then given).

Reference	$V_{\gamma,0}$, %	$YS_{0.2\%}$, MPa	UTS, MPa	TE, %	k
0.6C-1.5Si-890-250-16h	18	1448	1990	14.3	0.087
0.6C-1.5Si-890-220-22h	22	1246	2236 *	4.7 *	0.217
0.6C-1.5Si-950-250-22h	23	1404	1990	14.4	0.068
0.6C-1.5Si-950-220-40h	24	1295	2221	8.9	0.127
H0.6C-1.5Si-890-220-22h	21	1193	2158 *	5.8 *	0.195
1C-2.5Si-950-220-22h	43	1675	2185 *	3.9 *	0.203
1C-2.5Si-950-220-70h	33	1921	2277 *	5.7 *	0.181
1C-2.5Si-950-250-16h	37	1738	2106	16.8	0.058
1C-2.5Si-950-250-40h	35	1785	2101	15.8	0.078
1C-2.5Si-1050-220-40h	41	1768	2195 *	2.6 *	0.553
1C-2.5Si-1050-250-25h	34	1676	2088	14.9	0.048
1C-1.5Si-950-220-22h	40	1192	2063 *	3.0 *	1.009
1C-1.5Si-950-250-16h	33	1740	2170	10.7	0.130

$V_{\gamma,0}$ is the retained austenite content as measured in the grip; UTS is Ultimate tensile strength; TE is Total elongation; k is the constant in Equation (1), for a plastic strain expressed in %.

In an attempt to quantify the relationship between retained austenite stability and tensile ductility, the results were represented as per the following relationship [24]:

$$\ln(V_{\gamma,0}) - \ln(V_\gamma) \propto k\varepsilon_p \tag{1}$$

where $V_{\gamma,0}$ is the initial retained austenite content as measured in the specimen grip, and V_γ the retained austenite in the gauge length after application of a true plastic strain of ε_p. Some results are illustrated as an example in Figure 2. For all conditions investigated, the value of k was estimated using linear regression (throughout, this value is given for ε_p in percent).

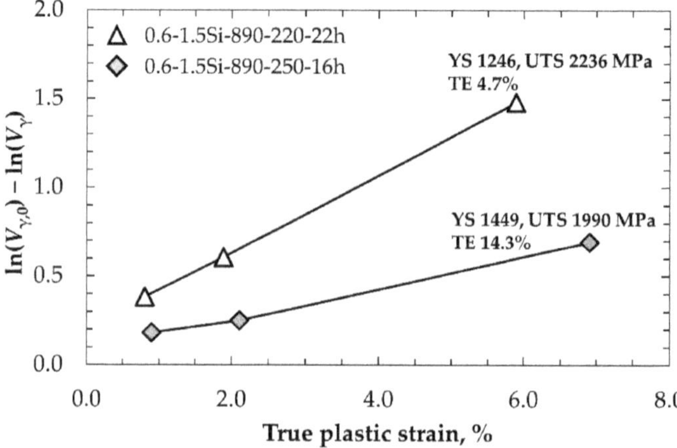

Figure 2. Destabilisation of retained austenite as a function of true plastic strain for two selected conditions.

Figure 3 shows the tensile elongation as a function of k for all conditions investigated. It is worth underlining that the corresponding dataset is for three different materials with a variety of heat-treatments. A first and clear correlation is that relating high values of k (rapid mechanical destabilisation of retained austenite with increasing strain) with brittle behaviour. Indeed, all conditions for which k values of more than ~0.2 were measured, led to brittle fracture during full tensile tests. Inversely, below that threshold, there appears to be a direct correlation between improved mechanical stability (as measured through k) and tensile ductility.

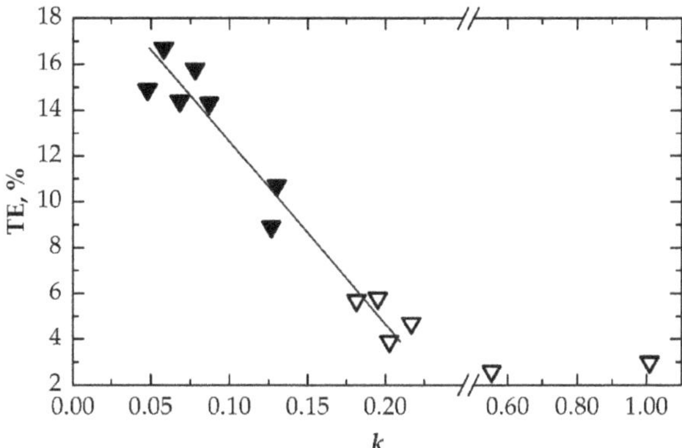

Figure 3. Tensile elongation as a function of the value of k for all conditions investigated. Hollow symbols are for brittle ruptures; full symbols are for ductile ruptures.

In addition to measurements carried out on specimens tested within the uniform elongation domain, retained austenite content was sometimes estimated on surfaces directly underneath the rupture surface using the first specimens having undergone full tensile tests. These measurements were associated with a very large texture uncertainty and are therefore to be taken with caution; they were nevertheless frequently below 10% (e.g., 4% for 1C-2.5Si-950-220-22h, 5% for 1C-2.5Si-950-250-40h). More reliably, a number of measurements within the uniform elongation domain yielded retained austenite content below 10% (0.6C-1.5Si-890-250-16h, γ_{res} 9% for 8% deformation; 0.6C-1.5Si-890-220-22h, γ_{res} 5% for 7% deformation).

4. Discussion and Conclusions

As discussed earlier, it has been proposed that the tensile ductility of bainite formed at low temperatures is correlated with the amount of retained austenite initially available, and limited to a percolation threshold, below which further plastic deformation is no longer possible [17].

The present data provide two important results. First, and as can be seen in Table 3, there is no correlation between initial retained austenite content and tensile ductility. In fact, retained austenite contents of 35%–40% can be associated with "brittle" behavior (1C-2.5Si-950-220-22h, 1C-2.5Si-1050-220-40h, 1C-1.5Si-950-220-22h) but also with very good tensile ductility (1C-2.5Si-950-250-16h). On the contrary, the approximate quantification of retained austenite stability via k exhibits an excellent correlation with tensile ductility. Furthermore, results suggest the existence of a critical value of k, beyond which brittle behavior cannot be avoided (low retained austenite stability). Below this value, the results suggest a continuous benefit in increasing retained austenite stability to enhance tensile ductility. In particular, the present results do not provide evidence that these microstructures may exhibit "excessive" austenite stability as suggested in earlier publications [22].

Second, the above results do not confirm the suggested existence of a threshold retained austenite content [17], below which ductile deformation is no longer possible.

Interestingly, poor ductility (high k values) was largely associated with transformation at 220 °C, whereas transformation at 250 °C tended to systematically provide ductile behavior (Table 3).

A detailed investigation (SEM, TEM, 3D-APT) of the potential origins of this difference will be published separately.

Acknowledgments: The authors gratefully acknowledge the support of the European Research Fund for Coal and Steel under grant number RFSR-CT-2012-00017.

Author Contributions: All contributed to the design of the experimental plan and the interpretation. Thomas Sourmail: general project definition, coordination, retained austenite measurements, analysis as per eq. 1. FG Caballero, CG Mateo, L Rivas-Morales and R. Rementeria: retained austenite measurements, discussion and interpretation R. Rementeria helped largely with manuscript formatting and figures. M. Kuntz: heat-treatments and tensile tests, interpretation.

Conflicts of Interest: The authors declare no conflict of interest.

References

1. Caballero, F.G.; Bhadeshia, H.K.D.H.; Mawella, K.J.A.; Jones, D.G.; Brown, P. Design of novel high strength bainitic steels: Part 1. *Mater. Sci. Technol.* **2001**, *17*, 512–516. [CrossRef]

2. Caballero, F.G.; Bhadeshia, H.K.D.H.; Mawella, K.J.A.; Jones, D.G.; Brown, P. Design of novel high strength bainitic steels: Part 2. *Mater. Sci. Technol.* **2001**, *17*, 517–522. [CrossRef]

3. Caballero, F.G.; Bhadeshia, H.K.D.H.; Mawella, K.J.A.; Jones, D.G.; Brown, P. Very strong low temperature bainite. *Mater. Sci. Technol.* **2002**, *18*, 279–284. [CrossRef]

4. Garcia-Mateo, C.; Caballero, F.G.; Bhadeshia, H.K.D.H. Development of hard bainite. *ISIJ Int.* **2003**, *43*, 1238–1243. [CrossRef]

5. Garcia-Mateo, C.; Caballero, F.; Bhadeshia, H. Low temperature bainite. *J. Phys. IV* **2003**, *112*, 285–288. [CrossRef]

6. Caballero, F.G.; Bhadeshia, H.K.D.H. Very strong bainite. *Curr. Opin. Solid State Mater. Sci.* **2004**, *8*, 251–257. [CrossRef]

7. Wang, T.S.; Li, X.Y.; Zhang, F.C.; Zheng, Y.Z. Microstructures and mechanical properties of 60Si2CrVa steel by isothermal transformation at low temperature. *Mater. Sci. Eng. A* **2006**, *438–440*, 1124–1127. [CrossRef]

8. Soliman, M.; Palkowski, H. Ultra-fine bainite structure in hypo-eutectoid steels. *ISIJ Int.* **2007**, *47*, 1703–1710. [CrossRef]

9. Timokhina, I.B.; Beladi, H.; Xiong, X.Y.; Adachi, Y.; Hodgson, P.D. Nanoscale microstructural characterization of a nanobainitic steel. *Acta Mater.* **2011**, *59*, 5511–5522. [CrossRef]

10. Garcia-Mateo, C.; Caballero, F.G.; Sourmail, T.; Kuntz, M.; Cornide, J.; Smanio, V.; Elvira, R. Tensile behaviour of a nanocrystalline bainitic steel containing 3 wt % silicon. *Mater. Sci. Eng. A* **2012**, *549*, 185–192. [CrossRef]

11. Sourmail, T.; Caballero, F.G.; Garcia-Mateo, C.; Smanio, V.; Ziegler, C.; Kuntz, M.; Elvira, R.; Leiro, A.; Vuorinen, E.; Teeri, T. Evaluation of potential of high si high c steel nanostructured bainite for wear and fatigue applications. *Mater. Sci. Technol.* **2013**, *29*, 1166–1173. [CrossRef]

12. Sourmail, T.; Smanio, V.; Ziegler, C.; Heuer, V.; Kuntz, M.; Caballero, F.G.; Garcia-Mateo, C.; Cornide, J.; Elvira, R.; Leiro, A.; et al. *Novel Nanostructured Bainitic Steel Grades to Answer the Need for High-Performance Steel Components (Nanobain)*; European Commission: Luxembourg, Luxembourg, 2013.

13. Garbarz, B.; Burian, W. Microstructure and properties of nanoduplex bainite–austenite steel for ultra-high-strength plates. *Steel Res. Int.* **2014**, *85*, 1620–1628. [CrossRef]

14. Soliman, M.; Palkowski, H. Development of the low temperature bainite. *Arch. Civ. Mech. Eng.* **2016**, *16*, 403–412. [CrossRef]

15. Garcia-Mateo, C.; Caballero, F.G.; Bhadeshia, H.K.D.H. Acceleration of low-temperature bainite. *ISIJ Int.* **2003**, *43*, 1821–1825. [CrossRef]

16. Tump, A.; Brandt, R. Graded high-strength spring-steels by a special inductive heat Treatment. *IOP Conf. Ser. Mater. Sci. Eng.* **2016**, *118*, 012021. [CrossRef]

17. Bhadeshia, H.K.D.H. Nanostructured bainite. *Proc. R. Soc. A* **2010**, *466*, 3–18. [CrossRef]

18. Sandvik, B.P.J.; Navalainen, H.P. Structure-property relationships in commercial low-alloy bainitic-austenitic steel with high strength, ductility, and toughness. *Met. Technol.* **1981**, *8*, 213–220. [CrossRef]

19. Bhadeshia, H.K.D.H. *Bainite in Steels. Transformations, Microstructure and Properties*, 2nd ed.; Institute of Materials, Minerals and Mining: London, UK, 2001.

20. Garcia-Mateo, C.; Caballero, F.G. The role of retained austenite on tensile properties of steels with bainitic microstructures. *Mater. Trans. JIM* **2005**, *46*, 1839–1846. [CrossRef]
21. Morales-Rivas, L.; Garcia-Mateo, C.; Kuntz, M.; Sourmail, T.; Caballero, F.G. Induced martensitic transformation during tensile test in nanostructured bainitic steels. *Mater. Sci. Eng. A* **2016**, *662*, 169–177. [CrossRef]
22. Garcia-Mateo, C.; Caballero, F.G.; Chao, J.; Capdevila, C.; de Andres, C.G. Mechanical stability of retained austenite during plastic deformation of super high strength carbide free bainitic steels. *J. Mater. Sci.* **2009**, *44*, 4617–4624. [CrossRef]
23. Avishan, B.; Garcia-Mateo, C.; Yazdani, S.; Caballero, F.G. Retained austenite thermal stability in a nanostructured bainitic steel. *Mater. Charact.* **2013**, *81*, 105–110. [CrossRef]
24. Sherif, M.Y.; Mateo, C.G.; Sourmail, T.; Bhadeshia, H. Stability of retained austenite in trip-assisted steels. *Mater. Sci. Technol.* **2004**, *20*, 319–322. [CrossRef]

Article

Evaluating Strengthening and Impact Toughness Mechanisms for Ferritic and Bainitic Microstructures in Nb, Nb-Mo and Ti-Mo Microalloyed Steels

Gorka Larzabal, Nerea Isasti, Jose M. Rodriguez-Ibabe and Pello Uranga *

CEIT and TECNUN, University of Navarra, 20018 San Sebastian, Basque Country, Spain; glarzabal@ceit.es (G.L.); nisasti@ceit.es (N.I.); jmribabe@ceit.es (J.M.R.-I.)
* Correspondence: puranga@ceit.es; Tel.: +34-943-212-800

Academic Editor: Carlos Garcia-Mateo
Received: 10 February 2017; Accepted: 17 February 2017; Published: 22 February 2017

Abstract: Low carbon microalloyed steels show interesting commercial possibilities by combining different "micro"-alloying elements when high strength and low temperature toughness properties are required. Depending on the elements chosen for the chemistry design, the mechanisms controlling the strengths and toughness may differ. In this paper, a detailed characterization of the microstructural features of three different microalloyed steels, Nb, Nb Mo and Ti-Mo, is described using mainly the electron backscattered diffraction technique (EBSD) as well as transmission electron microscopy (TEM). The contribution of different strengthening mechanisms to yield strength and impact toughness is evaluated, and its relative weight is computed for different coiling temperatures. Grain refinement is shown to be the most effective mechanism for controlling both mechanical properties. As yield strength increases, the relative contribution of precipitation strengthening increases, and this factor is especially important in the Ti-Mo microalloyed steel where different combinations of interphase and random precipitation are detected depending on the coiling temperature. In addition to average grain size values, microstructural heterogeneity is considered in order to propose a new equation for predicting ductile–brittle transition temperature (DBTT). This equation considers the wide range of microstructures analyzed as well as the increase in the transition temperature related to precipitation strengthening.

Keywords: microalloyed steels; niobium; molybdenum; titanium; mechanical properties; yield strength; impact toughness; modeling; microstructure; EBSD

1. Introduction

High Strength Low Alloy (HSLA) steels are widely used in the fabrication of beams, storage tanks, oil and gas pipelines, etc. In these applications, a balance between strength, toughness, weldability and cost is needed [1]. The careful design of the chemical composition of the steel in conjunction with an appropriate thermomechanical schedule can produce a wide variety of microstructures, from classical ferrite-pearlite combinations to more advanced non-polygonal/bainitic phases with an optimum balance of mechanical properties.

In this context, HSLA steels show a lower carbon content, which improves weldability and formability, but the lower mechanical properties that result from lower C contents can be counterbalanced by the addition of alloying elements such as Nb, Mo and Ti and an appropriate thermomechanical process. Each one of these elements affects different mechanisms. On one hand, many studies agree that Nb is able to induce strain accumulation in the austenite prior to transformation, providing significant microstructural refinement [2]. Mo, in addition to having a solute drag effect on the static recrystallization kinetics, enhances the formation of complex non-polygonal transformation products [3,4]. These strategies pursue finer final microstructures, which would result in better combinations of strength and toughness. On the other hand, steels microalloyed with Ti and Mo have an interesting combination of high strength and good formability because of the wide dispersion of nanometric sized titanium carbides within a fine matrix [5].

This study analyzes the relationship between microstructure and mechanical properties in three different microalloyed steels. Relevant microstructural features such as grain size refinement and fine precipitation can be controlled by the coiling temperature after hot rolling. Therefore, the potential of different strengthening mechanisms needs to be further explored in order to find the relationship between microstructure, process parameters (coiling temperature) and mechanical properties such as tensile and toughness properties. For that purpose, tensile and Charpy tests were performed for all the conditions. The contribution of the different strengthening mechanisms, such as grain size refinement, secondary phases, precipitation hardening, solid solution and dislocation strengthening have been calculated. A model that is able to predict the yield strength and impact transition temperature depending on the applied thermomechanical schedule and the chemical composition is described. The model is valid for the whole range of microstructures (ferritic and bainitic) and chemical compositions. The equation proposed for predicting the ductile–brittle transition temperatures (DBTT) also takes into account the influence of microstructural heterogeneity and secondary hard phases.

2. Materials and Methods

In the present study, three low carbon Nb, NbMo and TiMo microalloyed steels are selected. Their chemical composition is listed in Table 1.

Table 1. Chemical composition of the steels (weight percent).

Steel	C	Mn	Si	P	S	Ti	Nb	Mo	Al	N
Nb	0.040	1.55	0.20	0.017	0.006	-	0.034	-	0.01	0.005
NbMo	0.049	1.60	0.21	0.019	0.007	-	0.035	0.2	0.02	0.007
TiMo	0.048	1.61	0.20	0.020	0.006	0.09	-	0.2	0.02	0.004

Plane strain compression tests were performed at different simulated coiling temperatures (T_{coiling}, °C) following the thermomechanical schedule represented in Figure 1. The plane compression specimens were reheated at 1200 °C for 5 min, followed by a multipass deformation sequence. The first two deformations ($\varepsilon = 0.4$) at 1100 and 1000 °C were designed in order to ensure fine recrystallized austenite. Then, the specimens were deformed at 900 °C, below the non-recrystallized temperature, to obtain a deformed austenite prior to transformation. After the last deformation, the samples were cooled down at 10 °C/s to three different coiling temperatures (700, 600 and 500 °C), where the specimens were maintained for 90 min to simulate the coiling process. Finally, the samples were cooled down slowly (1 °C/s) to room temperature.

The strain distributes heterogeneously through the section of the plane strain compression specimen, as a result of friction and specimen/tool geometry. Therefore, the specimens used for the microstructural and mechanical (tensile and Charpy specimens) characterization were obtained from the central part of the plane strain compression specimens in order to minimize strain gradients. The microstructures were characterized after etching in 2% Nital via different characterization techniques: optical microscopy (OM, LEICA DMI5000 M, Leica Microsystems, Wetzlar, Germany) and field-emission gun scanning electron microscopy (FEGSEM, JEOL JSM-7000F, JEOL Ltd., Tokyo, Japan). In order to quantify the crystallographic features, electron backscattered diffraction (EBSD) scans were performed for all samples. For that purpose, the samples were polished to 1 µm, followed by a polish with colloidal silica. Orientation imaging microscopy was carried out on the Philips XL 30CP SEM with W-filament, using TSL (TexSEM Laboratories, Salt Lake City, UT, USA) equipment. Different scan step sizes were used depending on the resolution needed, varying from 0.1 µm for high resolution scans to 0.4 µm for unit size measurements. The total scanned area was about 200×200 µm^2. The study of the precipitation was performed using a transmission electron microscope (TEM, JEOL 2100, JEOL Ltd., Tokyo, Japan) with a voltage of 200 kV and a LaB_6 thermionic filament. Carbon extraction replicas and electropolished thin foils were used for this purpose.

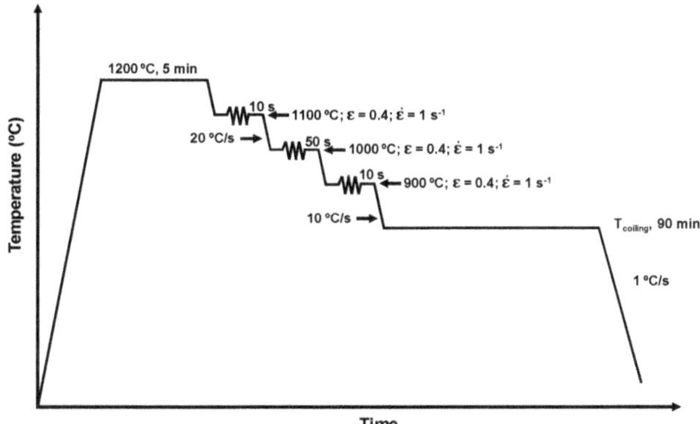

Figure 1. Schematic of the thermomechanical schedule.

Cylindrical tensile specimens with a gauge length of 17 mm and a diameter of 4 mm were machined from the plane strain compression samples. The tensile tests were performed at room temperature and with a strain rate of 10^{-3} s^{-1} on an Instron testing machine (Instron, Grove City, PA, USA) under strain control. The 0.2% proof stress and the ultimate tensile strength were determined as the mean value of two tests for each condition. Additionally, Charpy sub-size specimens (~4 × 10 × 55 mm^3) were machined and Charpy tests were performed (within an interval between −120 °C and 20 °C) in a Tinius Olsen Model Impact 104 pendulum impact tester with maximum capacity of 410 J. Specimens with a thickness of 4 mm are within the range of applicability of the proportionality rule [6]:

$$K_{v10} = \frac{10}{B} K_{vB} \tag{1}$$

where K_{v10} and K_{vB} are the impact energy for specimens that are 10 mm or B mm thick, respectively. The impact transition curves that are determined consider the modified hyperbolic tangent fitting algorithm proposed by Wallin [7]. Based on these curves, the temperature at which the sample shows a 50% ductile–brittle appearance transition temperature (DBTT) was calculated.

3. Results and Discussion

With the aim of obtaining a further understanding of the connection between chemical composition, coiling temperature, microstructure and mechanical properties, a detailed microstructural and mechanical property characterization was carried out. In the following sections, the evaluation of the influence of composition and coiling temperature on strength and toughness properties and the microstructural characterization results are described.

3.1. Microstructural Characterization

Both the tensile and toughness properties are controlled by different mechanisms such as solid solution hardening, microstructural refinement, dislocation hardening, precipitation hardening and secondary phases. With the aim of better understanding the role of those strengthening mechanisms in the yield strength and toughness, a deep microstructural analysis was carried out. In the subsequent sections, the results of this analysis are shown.

3.1.1. Microstructural Features, Unit Sizes and Homogeneity

The microstructures of the specimens obtained for the different coiling temperatures and composition were analyzed by optical (OM) and scanning electron microscopy (SEM) and the electron backscattering diffraction (EBSD) technique. Different transformation products can be distinguished depending on the chemical composition and the coiling temperature. In this article, the Iron and Steel Institute of Japan (ISIJ) Bainite Committee notation is adopted [2,8].

With regard to the Nb steel, when the highest coiling temperature of 700 °C is applied, a combination of Polygonal Ferrite (PF) and Pearlite (P) is observed. As the coiling temperature decreases, the mean size of the microstructure decreases, but no special change in microstructure was observed beyond the lack of pearlite at the coiling temperatures below 700 °C. On the other hand, in the NbMo and TiMo steels a microstructure composed of Polygonal Ferrite (PF), Pearlite (P) and Martensite/Austenite (MA) islands was obtained for the 700 °C coiling temperature. Reducing the coiling temperature leads to the formation of finer and more bainitic microstructures that are composed of PF, Quasipolygonal Bainite (QB), Granular Bainite (GB) and a small amount of MA islands, while in the NbMo steel the microstructure is composed of QB and GB and a higher fraction of MA islands. Conversely, at the lowest coiling temperature of 500 °C, a finer microstructure of QB and GB units is formed, and there is no trace of MA regions being retained between transformed phases in the NbMo and TiMo steels.

In addition to the qualitative microstructural analysis done by OM and SEM, an additional analysis using the EBSD technique was carried out due to the complexity of the analyzed microstructures. Low and high angle misorientation unit sizes were quantified, considering 2° and 15° misorientation criteria, respectively. As an example, Figure 2 shows Inverse Pole Figure maps (IPF) (Figure 2a,b) and grain boundary maps (Figure 2c,d) corresponding to the TiMo steel (Figure 2a,c) and the NbMo steel (Figure 2b,d) for a coiling temperature of 600 °C. The NbMo steel shows a finer microstructure when compared to the TiMo steel. Furthermore, a higher fraction of low angle boundaries (marked in red in Figure 2d) is noticed in the NbMo steel, reflecting the formation of more bainitic microstructures when Nb is added.

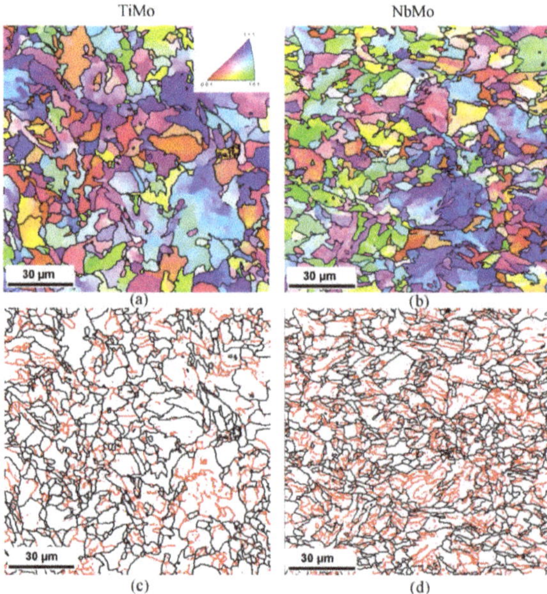

Figure 2. (**a,b**) IPF (Inverse Pole Figure) and (**c,d**) grain boundary maps obtained via EBSD in: TiMo (**a,c**); and NbMo (**b,d**) steels for a coiling temperature of 600 °C.

In Figure 3a,b, the mean unit sizes for low ($2°$, $D_{2°}$) and high ($15°$, $D_{15°}$) angle misorientation criteria are plotted. For both misorientation criteria, the mean unit size decreases as the coiling temperature decreases, and the Nb steel systematically presents the coarsest microstructure. On the other hand, the NbMo steel shows the smallest mean unit size for all coiling temperatures, the TiMo steel unit sizes being between the other two steels in all cases. Regarding the high angle criterion (Figure 3b), in the Nb and NbMo steels, there is an important decrease when the coiling temperature decreases from 700 to 600 °C. For example, in the NbMo steel, the mean unit size of the high angle criterion decreases from 6.8 to 4.4 μm. However, $D_{15°}$ remains approximately constant at lower coiling temperatures (4.4 μm for the coiling temperature (CT) of 600 °C and 4.8 μm for a CT of 500 °C). Although the same trend is observed for the Nb steel, slightly bigger unit sizes are attained. In contrast, in the TiMo steel, the decrease is more progressive. This effect can be explained by a more gradual modification of the microstructure in the TiMo steel from ferritic to bainitic microstructures. Similar trends are also observed for the low angle boundary unit sizes (Figure 3a).

In addition to mean size values, microstructural homogeneity is essential for toughness properties. In Figure 4a, unit size distributions are plotted taking into account the high angle misorientation criterion for the TiMo microalloyed steel and different coiling temperatures. A significant influence of coiling temperature is observable, where the distribution becomes more heterogeneous as the coiling temperature decreases (the wider tail of the distribution). Therefore, in order to quantify the effect of microstructural heterogeneity on ductile–brittle transition, a parameter that is able to evaluate the relevance of coarse grain fraction is required. A useful parameter for evaluating the length of the tail of a distribution is the relationship between the critical grain size known as $Dc_{20\%}$—which corresponds to the cutoff unit size at 80% area fraction in a grain size distribution—and the mean unit size, $D_{15°}$. The resulting parameter is defined as $Dc_{20\%}/D_{15°}$. Figure 4b shows the variation of the $Dc_{20\%}/D_{15°}$ parameter with coiling temperature. As a general trend, a progressive increase of the parameter as the coiling temperature is decreased is observed. This fact is related to the development of more non-polygonal or bainitic microstructural features as the coiling temperature decreases. This trend was previously reported both for coiling simulations as well as for continuously cooled samples [2,9].

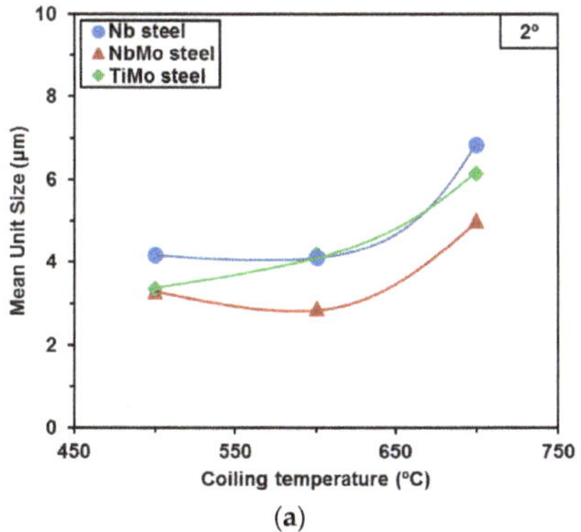
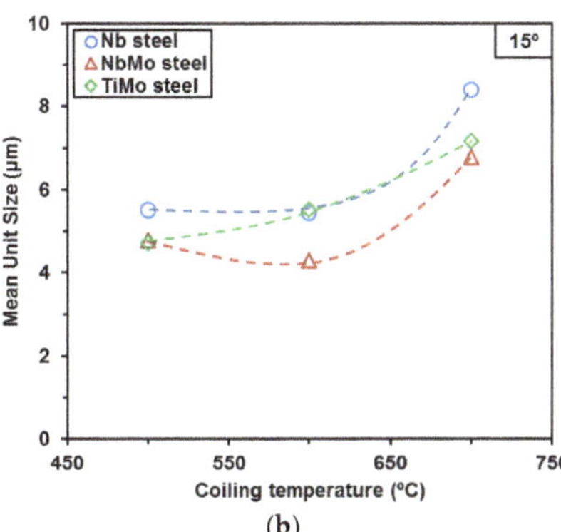

Figure 3. Influence of coiling temperature and chemical composition on the average unit size, using different threshold misorientation criteria: (**a**) low angle ($2°$); and (**b**) high angle ($15°$) boundaries.

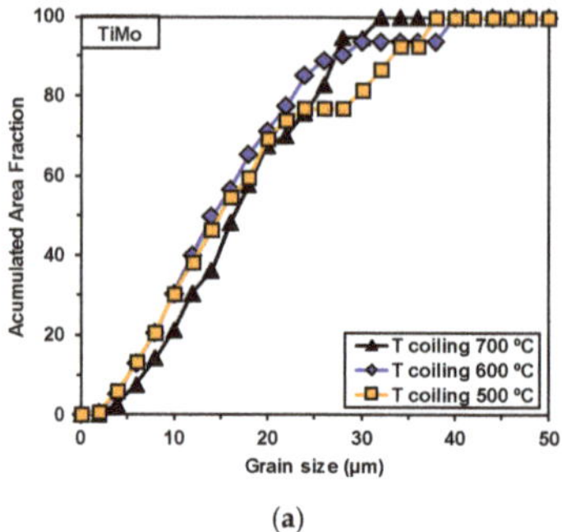

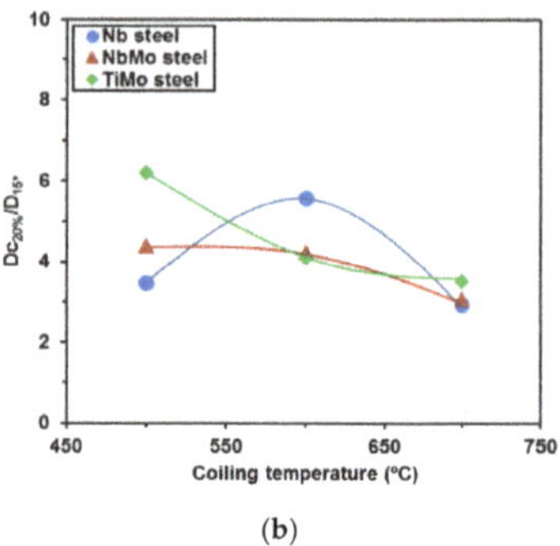

(a)	(b)

Figure 4. (**a**) Accumulated area fraction using a 15° misorientation criterion for the TiMo steel; and (**b**) evolution of the $Dc_{20\%}/D_{15°}$ parameter as a function of the coiling temperature.

3.1.2. Dislocation Density

In addition to quantifying unit sizes, strengthening due to dislocation networks has to be considered, as the progressive modification from ferritic to bainitic microstructures is relevant when coiling temperatures decrease. The dislocation density has been inferred using the Kernel Average Misorientation (KAM) parameter measured using EBSD. KAM reflects the local misorientation gradients within a given region. For that purpose, the approach proposed by Kubin and Mortensen is adopted (see Equation (2)) [10].

$$\rho = \frac{2KAM}{bu} \tag{2}$$

where b is the Burgers vector and u is the length related to the Kernel. The KAM for $\vartheta < 2°$ and the second neighbor distance have been predefined to calculate the parameter. In the current work, the length associated with the Kernel parameter (u) is 1.86 times the step size. With the aim of minimizing the frequent overestimation of the Kernel as determined by means of conventional EBSD, the measurements have been corrected using a recently proposed correlation between high-resolution EBSD and conventional EBSD for ferritic structures [11]. In Figure 5a,b, Kernel maps obtained after coiling at 600 °C are shown, for the TiMo and NbMo steel, respectively. Looking at the maps, higher KAM values are clearly observed in the NbMo steel than in the TiMo steel. KAM values of 0.97° and 0.76° are measured, for NbMo and TiMo steels, respectively, and resulting in dislocation densities of 1.76×10^{14} m^{-2} and 1.33×10^{14} m^{-2}. The formation of more bainitic phases in the NbMo steel results in this higher dislocation density. In Figure 5c, dislocation density values are plotted as a function of coiling temperature for all the steels studied. Dislocation density values vary from 6.20×10^{13} m^{-2} to 1.76×10^{14} m^{-2} depending on the chemical composition and coiling condition.

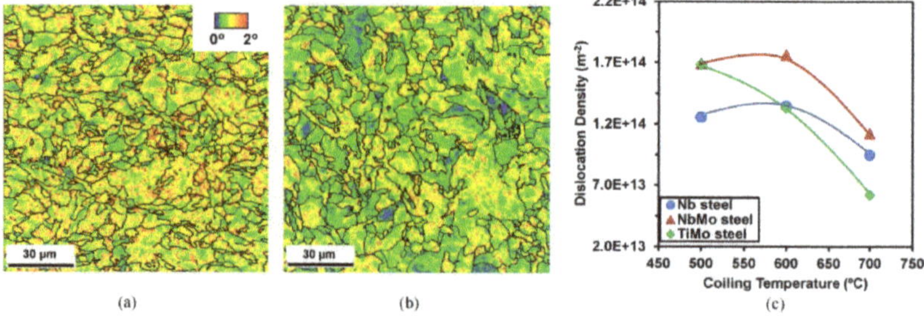

(a)	(b)	(c)

Figure 5. (**a,b**) Kernel maps obtained after a coiling temperature of 600 °C for NbMo and TiMo steels, respectively; and (**c**) dislocation density as a function of coiling temperature for all the microalloyed steels.

In the Nb steel, no considerable change is noticed for dislocation density, with approximately similar dislocation density values being measured in the entire range of coiling temperatures (increasing from 9.47×10^{13} m^{-2} to 1.34×10^{14} m^{-2} as the coiling temperature decreases from 700 to 600 °C). In the two steels containing Mo, a different trend is observable, where a more significant increment in dislocation density is detected as the coiling temperature decreases. For the TiMo steel, dislocation density increases progressively from 6.20×10^{13} m^{-2} to 1.68×10^{14} m^{-2} when the coiling temperature is decreased from 700 to 500 °C. The trends shown in Figure 5c confirm the formation of more bainitic phases characterized by the highest dislocation densities in the NbMo steel. The dislocation densities reported in the literature for different microstructures differ considerably. For ferritic phases, a wide range of dislocation density values have been proposed, from very low dislocation density of 10^{12} m^{-2} [12] to a higher ρ of 2.5×10^{14} m^{-2} [13]. In the present study, the microstructures observed after a coiling at 700 °C contain secondary phases (MA microconstituent and pearlite) apart from polygonal ferrite. Given that higher *KAM* values are measured close to those hard phases, higher dislocation density values are estimated compared to purely ferritic microstructures. In terms of the dislocation densities reported for bainitic microstructures, the measurements shown in the present study are within the range reported by other authors [14,15].

3.1.3. Precipitation

In order to evaluate how the different microalloying elements affect precipitation strengthening, a semi-quantitative analysis was performed. First, a preliminary study of fine precipitates was carried out on carbon extraction replicas, and the average precipitate size was measured in selected coiling temperatures and steels. The quantification was limited to particles smaller than 10 nm, which are considered effective for contributing to hardening. Figure 6a–c illustrates differences in precipitation after a coiling temperature of 700 °C in the Nb, NbMo and TiMo steels. Regarding the influence of microalloying elements on the precipitate density, the micrographs shown in Figure 6 suggest that the chemical composition strongly modifies the density of fine precipitates. Higher precipitate densities are achieved in the steels containing Mo (NbMo and TiMo). Furthermore, the addition of Mo leads to a refinement of the precipitates. Precipitate diameter decreases from 6.2 to 3.6 nm, for Nb and NbMo steel, respectively. These data are in line with previously reported results [9]. Several works reported that precipitates formed in ferrite became finer and denser as higher Mo contents are added [16]. The higher volume fraction of precipitates can be explained by the decrease in the driving force of carbide nucleation caused by Mo addition, as well as the increase in the density of nucleation sites. Wada and Pehlke [17] observed that Mo decreases both the driving force of precipitates and the diffusivity of the carbide-forming species. Akben [18] also confirmed that the addition of Mo can retard the precipitation of carbides. In reference to the effect of coiling temperature, Lee et al. [16] reported that the higher density of carbides is achieved as the coiling temperature decreases, since the high density of dislocations generated by the bainitic transformation could serve as nucleation sites for precipitates. Additionally, Isasti et al. [9] observed that the intermediate coiling temperature of 550 °C maximizes the strengthening effect of precipitation, due to the presence of a higher precipitate concentration. However, the authors reported that at lower coiling temperatures (450 °C) a less effective precipitation occurs.

With regard to the precipitation observed after coiling at 700 °C in the TiMo microalloyed steel (see Figure 6c), very fine Ti- and Mo-containing precipitates are detected in the extraction replicas. Moreover, it is clear that a higher fraction of fine precipitates is formed in TiMo steel. It is reported that the considerable strengthening effect observed in TiMo microalloyed steel is attributed to a superior coarsening resistance of the (Ti, Mo)C carbide as compared to other carbides such as TiC and (Ti, Nb)C [19,20]. Although carbon extraction replicas are suitable for measuring precipitate size distributions and the microanalysis of the precipitates, this method presents limitations for evaluating the distribution and volume fraction of precipitates. Therefore, characterization was completed by means of thin foils. TEM micrographs shown in Figure 7a,b indicate that interphase precipitation occurs after coiling at 700 °C in the TiMo steel. The size of precipitates is in the range of 2–10 nm. This precipitation is characterized by the formation of particles that form repeatedly in the austenite/ferrite interphase as the transformation front moves through the austenite. As shown in Figure 7a,b, these precipitates are arranged in sheets parallel to the instantaneous position of the austenite/ferrite interphase and the resulting microstructure is formed by numerous sheets of precipitates with a given spacing. A microanalysis of these fine precipitates indicates that they are Ti-enriched carbides with some residual Mo presence.

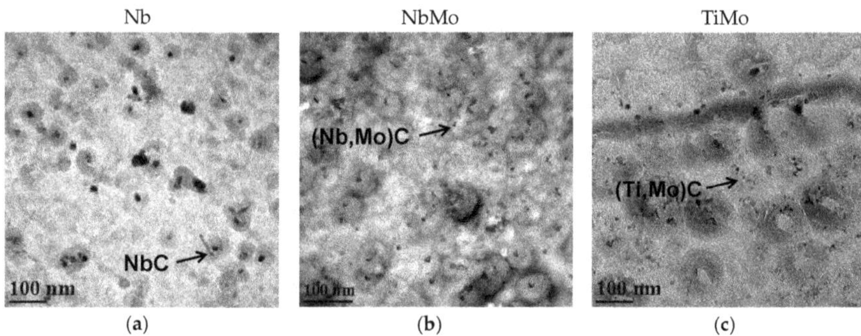

Figure 6. Presence of fine precipitates in the microstructures after coiling temperature of 700 °C for different steels: (**a**) Nb (NbC); (**b**) NbMo ((Nb, Mo)C); and (**c**) TiMo ((Ti, Mo)C).

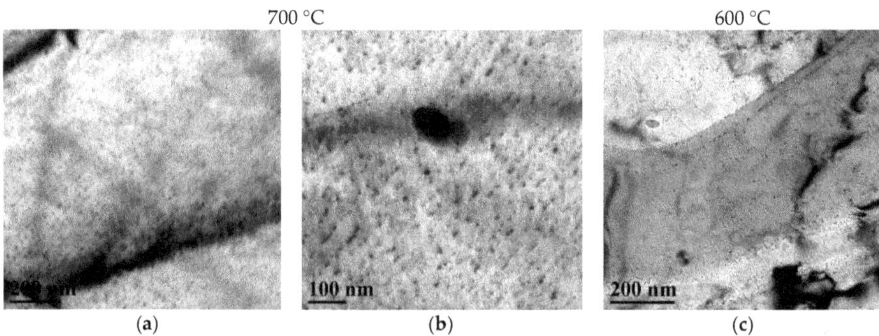

Figure 7. (**a**,**b**) Presence of interphase precipitation in the TiMo steel after coiling at 700 °C; and (**c**) random precipitation at 600 °C.

Differences in precipitation are observed depending on the coiling temperature. In the TEM micrographs presented in Figure 7, it is clear that the distribution of the precipitates varies as the coiling temperature decreases. When the highest coiling temperature is applied, interphase precipitation is detected, while at the intermediate coiling temperature of 600 °C a mixture between aligned precipitates and random precipitates is observed. Moreover, TiC rows are formed on the grain boundary. Chen [21] reported that compared with other ferrites, such as polygonal ferrite and massive ferrite, quasi-polygonal bainite has a relatively low yield ratio and high strain-hardening rates. The combination of quasi-polygonal bainite and dispersed nanoscale precipitates provided a favorable strengthening effect. Several studies support the fact that both interphase precipitation carbides and randomly precipitated carbides can coexist in the same specimen under various thermomechanical conditions [19,22]. After analyzing different precipitation times, Tamako et al. noticed that the ratio of randomly precipitated carbides to interphase precipitation carbides is no less than 0.8 when the steel sheet is isothermally held at 500 °C to 600 °C [23]. The differences between interphase and random precipitation have also been investigated by in-situ nanomechanical testers, evaluating the influence of both types of precipitation on the dislocation movement in ferrite grains and on nanomechanical properties [24]. A study carried out by Mukherjee et al. [25] also revealed the coexistence of nanoclusters and precipitate particles. They determined that a bimodal distribution of larger (8–10 nm) precipitates coexisted with smaller nanoclusters (3 nm) within the interphase sheets/rows.

3.1.4. Secondary Phases

Besides the effect of the mentioned microstructural aspects, the presence of secondary phases, such as pearlite and MA islands, affects both tensile and impact toughness properties. Therefore, it is necessary to quantify the fraction and the size of these hard phases. As mentioned previously, the presence of MA islands is limited to the steels containing Mo (TiMo and NbMo) and high coiling temperatures (700 and 600 °C). Concerning pearlite, all the steels show this constituent at the highest coiling temperature of 700 °C. Both the morphology and size of the secondary phase differs considerably depending on the coiling temperature. In Table 2, the mean size of the MA islands, as well as the fraction of secondary phases, such as MA microconstituent and pearlite, are summarized. In the NbMo steel, as the coiling temperature decreases, non-polygonal phases are promoted and the diffusion of C is reduced, decreasing the C concentration in the non-transformed austenite [26]. These mechanisms lead to the refinement of MA islands. Mean sizes ranging from 2.5 to 1.2 µm are measured for the coiling temperature of 700 and 600 °C, respectively. A combination of MA (2.3%) and pearlite (13.9%) is detected after coiling at 700 °C, while the formation of pearlite is eliminated at the intermediate coiling temperature of 600 °C (see Table 2). In the TiMo steel, no significant variation in MA size is observed, which is associated with the formation of polygonal phases in the range between 700 and 600 °C. For the lowest coiling temperature of 500 °C, due to the significant reduction of the diffusion of carbon, MA islands cannot be identified in both steels.

Table 2. Mean size of MA islands, MA fraction and pearlite fraction for each steel and coiling temperature.

Steel	$T_{coiling}$ (°C)	Mean Size of MA Islands (µm)	MA Fraction (%)	Pearlite Fraction (%)
Nb	700	0	0	5.3
NbMo	700	2.5	2.3	13.9
	600	1.2	2.8	0
TiMo	700	1.9	1.7	3.9
	600	2.1	1.0	0

3.2. Mechanical Behavior

3.2.1. Tensile Properties

Tensile data (yield and tensile strength) are plotted in Figure 8a,b as a function of coiling temperature for all the microalloyed steels. The overall tensile property parameters are listed in Table 3. Looking at the evolution of yield strength (see Figure 8a), three strength levels can be clearly distinguished depending on the composition of the steel. The highest strength level is attained in TiMo steel, followed by the NbMo and Nb steels. A similar trend is observed for tensile strength (see Figure 8b). The addition of Mo to Nb microalloyed steels promotes an increase in the tensile properties (yield and tensile strength) mainly at intermediate (600 °C) and low coiling temperatures (500 °C). The addition of Mo induces the formation of non-polygonal phases, and this microstructural modification promotes an increment in strength through a substructure formation, an increase of the dislocation density and a higher presence of fine precipitates. These results are in line with previous reports [4]. This effect can additionally be associated with the presence of a higher fraction of fine precipitates at an intermediate coiling temperature (600 °C) in both Mo-containing steels (NbMo and TiMo). However, at the lowest coiling temperature of 500 °C, the strength decreases significantly (see Figure 8). This tensile improvement at intermediate temperatures can be justified by the presence of nanoscale precipitates distributed in the ferrite or bainitic matrix.

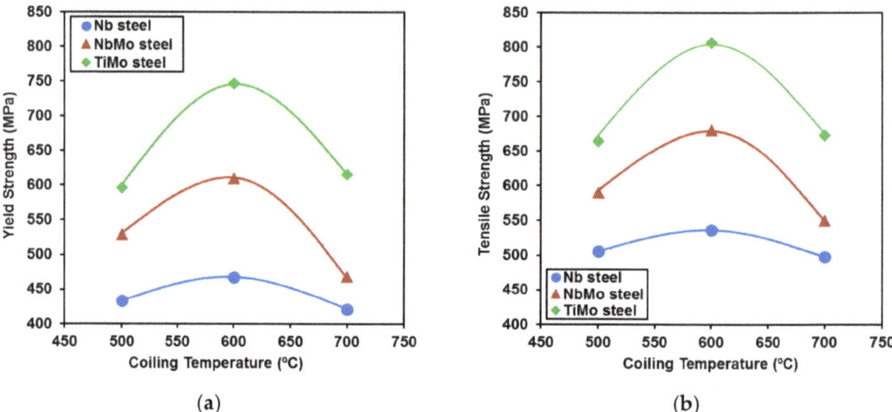

Figure 8. Evolution of the: yield strength (**a**); and tensile strength (**b**) as a function of coiling temperature.

Table 3. Yield Strength (*YS*), Tensile Strength (*TS*), Elongation (%) and Area Reduction (%) values for the TiMo, NbMo and Nb steels and coiling temperatures of 700 °C, 600 °C and 500 °C.

Steel	T_{coiling} (°C)	YS (MPa)	TS (MPa)	Elongation (%)	Area Reduction (%)
	700	421	498	39	87
Nb	600	468	536	33	89
	500	434	506	39	89
	700	468	550	28	82
NbMo	600	610	681	26	85
	500	529	590	33	90
	700	616	674	29	82
TiMo	600	747	807	27	79
	500	597	665	26	84

The maximum tensile properties (yield strength and tensile strength) are reached at the intermediate coiling temperature of 600 °C in TiMo and NbMo steel (see Figure 8 and Table 3). For the TiMo steel, maximum yield and tensile strengths of 747 and 807 MPa are obtained. Funakawa [5] reported the potential to improve strength in TiMo microalloyed steels through interphase precipitation. They proposed a strength improvement of 300–350 MPa, which is 2–3 times higher than that expected from conventional random precipitation hardening in microalloyed steels. Other authors provide support for the maximum tensile properties being reached at intermediate coiling temperatures in the range between 600 and 630 °C [27]. In terms of elongation (see Table 3), the Nb steel shows the highest values. The NbMo steel shows some dependency on MA presence, as the elongation drops sharply for the 700 °C coiling temperature, where big MA islands are formed. For the TiMo steel, however, the elongation values remain constant at approximately 27% even for yield strength values as high as 747 MPa. This suggests that the combination of a mostly ferritic matrix and very fine titanium carbides is a suitable route for combining high strength and good cold formability using a combination of titanium and molybdenum.

3.2.2. Toughness Properties

The impact transition curves corresponding to the different compositions and a coiling temperature of 700 °C are plotted together in Figure 9a. In terms of the effect of chemical composition, the best toughness properties are achieved for the Nb steel, followed by the NbMo and finally the TiMo steel. For example, for the 700 °C coiling temperature, 50% ductile–brittle appearance transition temperature (DBTT) values of −109, −94 and −51 °C are measured for the Nb, NbMo and TiMo steels, respectively. In the Nb and NbMo steels the transition from the ductile regime to the brittle one is produced in a narrow temperature range (the transition curve is almost vertical in Nb steel), whereas in the TiMo steel this transition temperature range is wider (see Figure 9a).

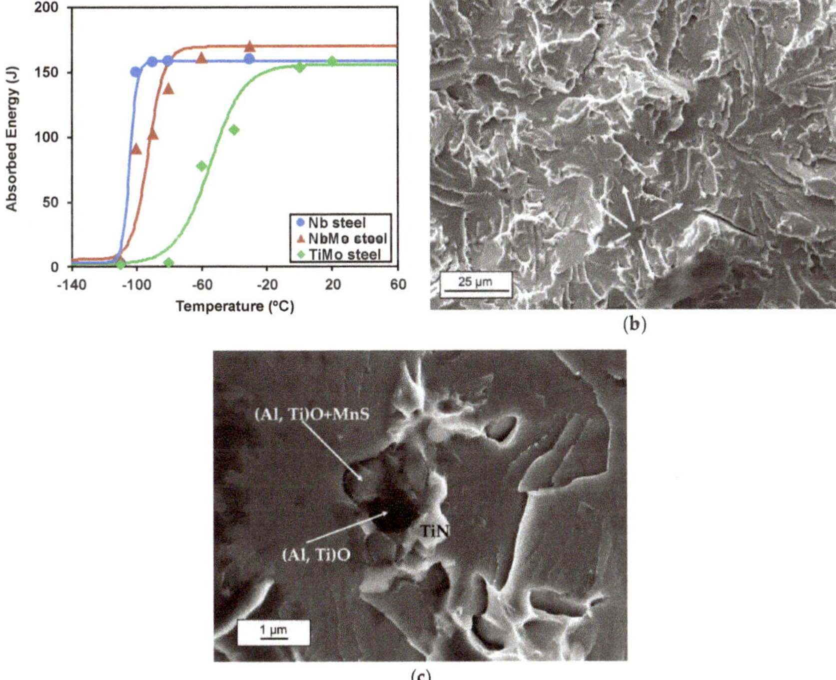

Figure 9. (**a**) Absorbed energy as a function of test temperature with a coiling temperature of 700 °C. (**b**,**c**) Fracture images corresponding to the TiMo steel and a coiling temperature of 700 °C, showing the origin of the brittle fracture.

Detailed fractographic examination was carried out on the tested Charpy samples for the purpose of classifying and evaluating possible cleavage crack-initiation sites and microstructural features in their vicinity. In Figure 9b,c, a cleavage initiation site is shown for the TiMo steel coiled at 700 °C at different magnifications. This case corresponds to a cleavage-initiation site due to the presence of a complex inclusion formed by TiN, (Mn, Ti)S and oxides. Nevertheless, a fractography examination carried out in NbMo fracture surfaces shows that the initiation appears in a secondary phase with a morphology typical of the MA constituent, which is in agreement with previously published works [28]. In the Nb steel, crack initiators are not easily detected, but in etched samples the presence of carbides seems to be the most important factor in controlling the crack initiation.

3.3. Predicting Mechanical Properties

3.3.1. Yield Strength Prediction

The yield strength of low carbon microalloyed steels can be described as a combination of different strengthening contributions. Although the most widely used approach is based on a linear summation of the contributions [29–31], several non-linear relationships, which consider the interaction of different strengthening mechanisms, have also been reported in the literature [27,32]. In the present study, a linear approach based on the sum of the contributions (solid solution [33], grain size [15], dislocations [10], presence of secondary phases [34] and fine precipitation [35]) has been considered (see Equation (3)). To estimate the individual contributions, equations previously reported in the literature have been employed and are listed in Equations (4)–(8) (see Nomenclature for symbol description). A more detailed description of the expressions can be found in [36]. For this study, due to the lack of an accurate measurement of precipitate volume fraction, the contribution of fine precipitation (σ_{ppt}) was estimated by subtracting the strengthening associated with all the other contributions from the experimental yield strength.

$$\text{General}: \ \sigma_y = \sigma_0 + \sigma_{ss} + \sigma_{gs} + \sigma_\rho + \sigma_{MA} + \sigma_{ppt} \tag{3}$$

$$\text{Solid solution}: \ \sigma_{ss} = \sigma_0 + 32.3Mn + 83.2Si + 11Mo + 354\left(\%N_{free}\right)^{0.5} \tag{4}$$

$$\text{Grain size}: \ \sigma_{gs} = 1.05\alpha M\mu\sqrt{b}\left[\sum_{2\leq\theta_i\leq15^\circ} f_i\sqrt{\theta_i} + \sqrt{\frac{\pi}{10}}\sum_{\theta_i\geq15^\circ} f_i\right] \cdot D_{2^\circ}^{\ -0.5} \tag{5}$$

$$\text{Dislocations}: \ \sigma_\rho = \alpha M\mu b\sqrt{\rho}, \ \text{where } \rho = \frac{2\vartheta}{ub} \tag{6}$$

$$\text{MA islands}: \ \sigma_{MA} = 900 f_{MA} \tag{7}$$

$$\text{Precipitation}: \ \sigma_{ppt} = 10.8\frac{f_v^{\ 0.5}}{x}\ln\left(\frac{x}{6.125\times10^{-4}}\right) \tag{8}$$

According to the Equations (3)–(8), the influence of each strengthening mechanism in yield strength has been plotted in Figure 10a–c as a function of coiling temperature. Concerning the Nb and NbMo steels, Figure 10a,b illustrates that the most predominant strengthening mechanism is related to grain size refinement, obtaining contributions from 203 to 312 MPa. It is clearly noticed that adding Mo (in Nb steel) and reducing the coiling temperature led to the increment of the term related to grain size. With regard to the influence of the strengthening contribution due to dislocation density, Figure 10a,b shows that this term increases as the coiling temperature decreases as it is associated with the formation of more non-polygonal or bainitic phases. Therefore, the NbMo steel shows a slightly higher effect relative to dislocations when compared to the Nb steel. This could be explained by the effect of Mo in the delay of phase transformations, providing the formation of more bainitic phases with a bigger dislocation density. In terms of the effect of fine precipitation, no significant contribution is observed in the Nb microalloyed steel in the entire range of coiling temperature. It is worth emphasizing that, in both Nb-based grades, the amount of this element is relatively too low (~0.035%) to have an important role in precipitation hardening. In the 500 °C coiling temperature prediction, the model overestimates the experimental values by 49 MPa by adding the measured contribution (the experimental value is indicated in Figure 10a). When Mo is added, a slightly higher contribution of precipitates is observable, reaching a maximum contribution of 50 MPa at the intermediate coiling temperature of 600 °C.

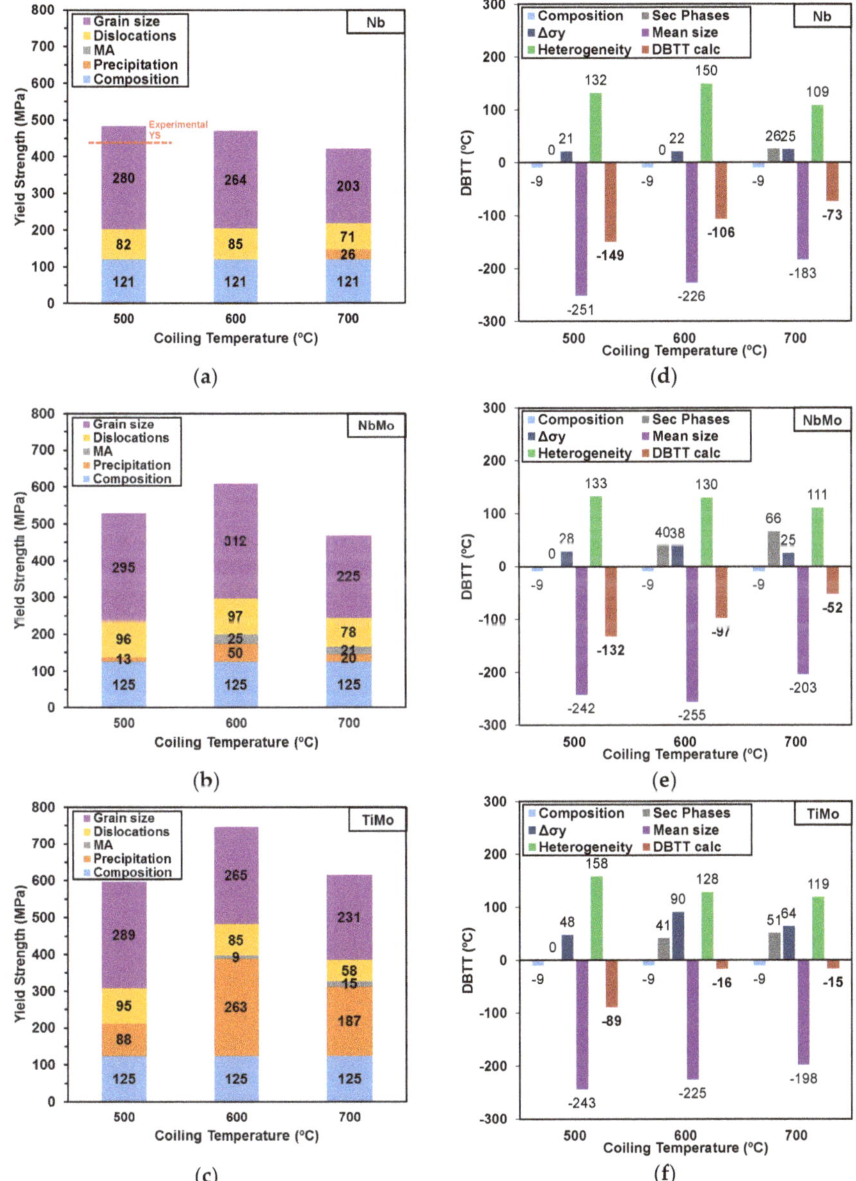

Figure 10. (**a**–**c**) Individual contributions (grain size, dislocations, MA islands, precipitation and composition) weighted according to Equation (3) for each coiling temperature. (**d**–**f**) Estimated contributions to the DBTT as a function of the coiling temperature and for all the microalloyed steels: (**a**,**d**) Nb; (**b**,**e**) NbMo; and (**c**,**f**) TiMo.

The trends observed in the TiMo microalloyed steel differ noticeably. In this case, the two most important strengthening contributions are due to grain size and fine precipitation. Even though similar terms related to grain size are shown in Figure 10c, a considerably higher precipitation contribution has been estimated for the TiMo microalloyed steel in all the coiling temperatures. Additionally, it is observed that the highest precipitation contribution is reached at the coiling temperature of 600 °C (263 MPa), where the maximum yield strength value of 747 MPa was measured. Similar precipitation contributions in the range of 220–320 MPa have been recently reported in other studies, defining 620 °C as the optimal coiling temperature, at which the best combination between tensile and toughness properties are achieved [37]. Nevertheless, at the lowest coiling temperature of 500 °C, precipitation hardening drops considerably from 263 MPa to 88 MPa. Other works also report less effective precipitation hardening at low coiling temperatures [36,38].

3.3.2. DBTT Prediction

Several relationships have been proposed in the literature for predicting impact transition temperatures [33,39]. In a recent study on low carbon Nb and NbMo microalloyed steels [28], the classical empirical equation proposed by Pickering and Gladman [33] was modified in order to take into account the effect of microstructural heterogeneity and the presence of hard secondary phases such as MA islands.

In the present study, higher transition temperatures were experimentally measured when compared to predictions calculated with the equation proposed in Reference [28]. This underestimation of DBTT values is mainly relevant for the TiMo steel, due to a significantly higher contribution of the term related to fine precipitation and dislocation density ($\Delta\sigma_y = \sigma_\rho + \sigma_{ppt}$), which is considerably lower for the Nb and NbMo steels. The increment in transition temperature by the increase in $\Delta\sigma_y$ is reported as having different values, depending on the source. Gladman et al. [40] reported an increase in yield strength of 0.35 °C·MPa^{-1}, whereas Gray [41] reported a value of 0.55 °C·MPa^{-1}. Pickering [42] noted different effects of $\Delta\sigma_y$ according to microstructure morphology. Pickering [42] noticed that a lower increase in yield strength is obtained for a bainitic microstructure (or acicular ferrite) than for a polygonal ferritic microstructure. An effect of 0.45 °C·MPa^{-1} is observed by Pickering for a ferritic phase, while this effect is reduced to 0.26 °C·MPa^{-1} for a bainitic one [42]. In the current study, in order to minimize the underestimation of the impact transition temperature, the detrimental contribution of $\Delta\sigma_y$ to DBTT has been set at 0.26 °C·MPa^{-1} for all the steels and microstructures analyzed. By modifying the weight of the term related to precipitation and dislocations, the definition of new prefactors for the MA island size and heterogeneity terms were fitted from the experimental DBTT temperatures and resulted in values equal to 18 and 63, respectively. The modified equation is able to predict impact transition temperatures (DBTT), and shown in Equation (9):

$$\begin{aligned} \text{DBTT (°C)} = \quad & -11\text{Mn} + 42\text{Si} + 700(\text{N}_{\text{free}})^{0.5} + 15(\text{pct Pearlite} + \text{pct MA})^{\frac{1}{3}} \\ & +0.26\Delta\sigma_y - 14(\text{D}_{15°})^{-0.5} + 63\left(\frac{\text{D}_{c20\%}}{\text{D}_{15°}}\right)^{0.5} + 18(\text{D}_{\text{MA}})^{0.5} - 42 \end{aligned} \tag{9}$$

The first two terms in Equation (9) are related to the solid solution contribution. Nitrogen is considered to be precipitated by forming the undissolved and strain-induced niobium carbonitrides, and in the case of the TiMo steel, all nitrogen is tied by titanium due to its hyperstoichiometric composition. Consequently, free nitrogen is considered to be zero at room temperature for all the chemical compositions. The percent of secondary phases is the sum of the fraction of pearlite and MA islands. $\text{D}_{15°}$ refers to the effective cleavage unit size (see Figure 3b) and D_{MA} is the size of the formed MA islands. Finally, the influence of heterogeneity is included by introducing the $\text{Dc}_{20\%}/\text{D}_{15°}$ factor (Figure 4). The comparison between the experimental DBTT values and the prediction calculated by Equation (9) is shown in Figure 11. The results obtained by Isasti et al. [28] have been included in the graph and computed to fix the constants. An accurate estimation of DBTT values is obtained for all the steels and different coiling temperatures studied.

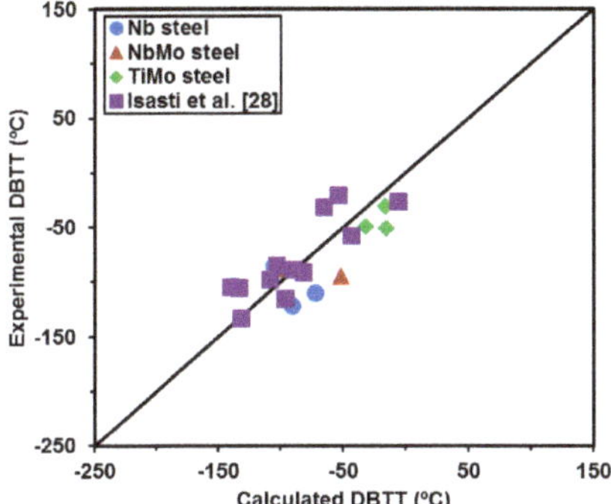

Figure 11. Comparison between experimental and calculated DBTT values predicted by Equation (9). The results obtained by Isasti et al. [28] have been included.

Regarding toughness, the values of the individual contribution to DBTT are represented in Figure 10d–f for each chemical composition (Nb, NbMo and TiMo steels, respectively) and the entire range of coiling temperatures. The predicted impact transition temperature according to Equation (9) has also been plotted in red. It is clear that the contribution related to grain size is the most relevant term, and in conjunction with composition, they are the only mechanisms that enhance the toughness properties. The term associated with unit size refinement ranges between 183 and 251 °C. The fine precipitates and dislocation density (included in the $\Delta\sigma_y$ term) result in an increment in the DBTT temperature. Regarding $\Delta\sigma_y$, the term varies considerably depending on the chemical composition and coiling temperature. The lowest $\Delta\sigma_y$ values are measured for the Nb steel, followed by the NbMo and TiMo steels. Moreover, the results plotted in Figure 10 suggest that the maximum precipitation + dislocation term is reached at the intermediate coiling temperature of 600 °C for both steels containing Mo (see Figure 10b,c). For example, in the TiMo steel, the contribution associated with $\Delta\sigma_y$ is about 64, 90 and 48 °C, after a coiling at 700, 600 and 500 °C, respectively. This trend could be attributed to more effective precipitation at the intermediate coiling temperature of 600 °C relative to the other coiling temperatures.

In addition to grain size, composition and the increase in yield strength, the contribution of heterogeneity as well as secondary phases has to be taken into account. Both contributions cause an increment in DBTT. In Figure 10, the contribution related to heterogeneity varies from 109 to 158 °C and increases as the coiling temperature decreases. Finally, the term concerning secondary phases includes a fraction of pearlite, a fraction of MA microconstituent and a mean size of MA islands. It is observed that the detrimental effect of the secondary phase is higher in the NbMo and TiMo steels than in the Nb steel, mainly at the highest coiling temperature of 700 °C. This contribution might cause a shift of 40–60 °C. Obviously, steel cleanliness is a key factor in controlling toughness properties. Different combinations of titanium oxides and nitrides were detected (see Figure 9) as trigger points for brittle fracture initiation.

4. Conclusions

A complete microstructural characterization procedure using EBSD and TEM is described to quantitatively evaluate the different mechanisms affecting yield strength and impact toughness. The proposed analysis and quantification methods are valid for the whole range of chemical composition and microstructures analyzed.

The quantification of each strengthening contribution has shown that crystallographic unit sizes play a major role up to 50%–70% of the yield strength, especially for the Nb and NbMo steels. Precipitation strengthening, together with average grain size, becomes predominant for the TiMo steel. Very fine interphase precipitation was detected in the TiMo showing a high volume fraction. The distribution of the nanoprecipitates varies from interphase row precipitates formed during austenite to ferrite decomposition to more randomly distributed particles within a non-polygonal matrix. The combination of Nb and Mo in the NbMo steel reduces the precipitate sizes and therefore increases the strengthening contribution of fine carbides precipitated at intermediate temperatures, showing a maximum for the intermediate coiling temperature of 600 °C. In this case, the spatial distribution is more random, as the transformed matrix is composed of granular and quasipolygonal bainite.

A study of the influence of coiling temperatures and chemical composition on impact toughness was also performed. In addition to mechanisms such as mean crystallographic unit sizes, secondary phase fractions, precipitation and the effect of dislocation density strengthening on transition temperatures, the impact of microstructural heterogeneity and MA island sizes has been expanded from a previously reported equation. The modified model for predicting the ductile–brittle transition temperature shows that the effect of precipitation strengthening on impact toughness when very high values are reached, such as the ones in the TiMo steel, is lower than $0.5 \cdot \Delta\sigma_y$ and it has been fixed at $0.26 \cdot \Delta\sigma_y$, which is valid for the three steels.

Acknowledgments: The authors acknowledge a research grant from the European Commission Research Fund for Coal and Steel (RFSR-CT-2013-00007) as well as the financial support of the Basque Government (PI-2014-1-129).

Author Contributions: Gorka Larzabal carried out the experiments and wrote the manuscript; Nerea Isasti analyzed the data and wrote the manuscript; Jose M. Rodriguez-Ibabe supervised the results and edited the manuscript; Pello Uranga managed the project and edited the manuscript.

Nomenclature

σ_y	Yield Strength
σ_0	Lattice friction stress
σ_{ss}	Strengthening contribution due to solid solution
σ_{gs}	Strengthening contribution due to grain size
σ_ρ	Strengthening contribution due to dislocations
σ_{ppt}	Strengthening contribution due to precipitation
σ_{MA}	Strengthening contribution due to MA islands
α	Numerical factor
M	Taylor factor
μ	Shear modulus
b	Burgers vector magnitude
f_i	Relative frequency
θ_i	Misorientation angle in the interval i
$D_{2°}$	Mean unit size using the 2° low angle boundary criterion
ρ	Dislocation density
ϑ	Kernel average misorientation
u	Unit length related to Kernel
f_{MA}	Volume fraction of MA islands
f_v	Volume fraction of precipitates
x	Mean planar intercept diameter of the particles

References

1. Jansto, S.G. Niobium-bearing steel development for value-added structural applications. In *New Developments on Metallurgy and Applications of High Strength Steels*; TMS: Warrendale, PA, USA, 2008; pp. 1313–1326.

2. Isasti, N.; Jorge-Badiola, D.; Taheri, M.L.; Uranga, P. Phase transformation study in Nb-Mo microalloyed steels using dilatometry and EBSD quantification. *Metall. Mater. Trans. A* **2013**, *44*, 3552–3563. [CrossRef]

3. Huang, B.M.; Yang, J.R.; Huang, C.Y. The synergistic effect of niobium-molybdenum additions on the microstructure of low-carbon bainitic steel. In *Fundamentals and Applications of Mo and Nb Alloying in High Performance Steels*, 2nd ed.; Mohrbacher, H., Ed.; CBMM, IMOA and TMS: Liège, Belgium, 2015; Volume 2, pp. 29–51.

4. Cizek, P.; Wynne, B.P.; Davies, C.H.J.; Hodgson, P.D. The Effect of Simulated Thermomechanical Processing on the Transformation Behavior and Microstructure of a Low-Carbon Mo-Nb Linepipe Steel. *Metall. Mater. Trans. A* **2015**, *46*, 407–425. [CrossRef]

5. Funakawa, Y.; Shiozaki, T.; Tomita, K.; Yamamoto, T.; Maeda, E. Development of High Strength Hot-rolled Sheet Steel Consisting of Ferrite and Nanometer-sized Carbides. *ISIJ Int.* **2004**, *44*, 1945–1951. [CrossRef]

6. Mintz, B.; Peterson, G.; Nassar, A. Structure-property relationships in ferrite-perlite steels. *Ironmak. Steelmak.* **1994**, *21*, 215–222.

7. Wallin, K. *Modified Tank Fitting Algorithm for Charpy Impact Data*; Research Seminar on Economical and Safe Application of Modern Steels for Pressure Vessels: Aachen, Germany, 2003.

8. Araki, T.; Kozasu, I.; Tankechi, H.; Shibata, K.; Enomoto, M.; Tamehiro, H. *Atlas for Bainitic Microstructures*; ISIJ: Tokyo, Japan, 1992; Volume 1.

9. Isasti, N.; Jorge-Badiola, D.; Taheri, M.L.; Uranga, P. Microstructural and Precipitation Characterization in Nb-Mo Microalloyed Steels: Estimation of the Contributions to the Strength. *Met. Mater. Int.* **2014**, *20*, 807–817. [CrossRef]

10. Kubin, L.P.; Mortensen, A. Geometrically necessary dislocations and strain-gradient plasticity: A few critical issues. *Scr. Mater.* **2003**, *48*, 119–125. [CrossRef]

11. Isasti, N.; Badiola, D.J.; Alkorta, J.; Uranga, P. Analysis of Complex Steel Microstructures by High-Resolution EBSD. *JOM* **2016**, *68*, 215–223. [CrossRef]

12. Roberts, M.J. Effect of transformation substructure on the strength and toughness of Fe-Mn alloys. *Metall. Trans.* **1970**, *1*, 3287–3294.

13. Wang, R.; García, C.I.; Hua, M.; Cho, K.; Zhang, H.; Deardo, A.J. Microstructure and Precipitation Behavior of Nb, Ti Complex Microalloyed Steel Produced by Compact Strip Processing. *ISIJ Int.* **2006**, *46*, 1345–1353. [CrossRef]

14. Garcia-Mateo, C.; Caballero, F.G.; Capdevilla, C.; de Andres, C.G. Estimation of dislocation density in bainitic microstructures using high-resolution dilatometry. *Scr. Mater.* **2009**, *61*, 855–858. [CrossRef]

15. Iza-Mendia, A.; Gutiérrez, I. Generalization of the existing relations between microstructure and yield stress from ferrite–pearlite to high strength steels. *Mater. Sci. Eng. A* **2013**, *561*, 40–51. [CrossRef]

16. Lee, W.B.; Hong, S.G.; Park, C.G.; Park, S.H. Carbide Precipitation and High-Temperature Strength of Hot-rolled High-Strength, Low-Alloy Steels Containing Nb and Mo. *Met. Mater. Trans. A* **2002**, *33*, 1689–1698. [CrossRef]

17. Wada, H.; Pehlke, R.D. Nitrogen solubility and nitride formation in austenitic Fe-Ti alloys. *Metall. Trans. B* **1985**, *16*, 815–822. [CrossRef]

18. Akben, M.G.; Bacroix, B.; Jonas, J.J. Effect of Vanadium and Molybdenum addition on High Temperature Recovery, Recrystallization and Precipitation Behavior of Niobium-based Microalloyed Steels. *Acta Metall.* **1983**, *31*, 161–174. [CrossRef]

19. Chen, C.Y.; Yen, H.W.; Kao, F.H.; Li, W.C.; Huang, C.Y.; Yang, J.R.; Wang, S.H. Precipitation hardening of high-strength low-alloy steels by nanometer-sized carbides. *Mater. Sci. Eng. A* **2009**, *499*, 162–166. [CrossRef]

20. Wang, Z.; Zhang, H.; Guo, C.; Liu, W.; Yang, Z.; Sun, X.; Zhang, Z.; Jiang, F. Effect of molybdenum addition on the precipitation of carbides in the austenite matrix of titanium micro-alloyed steels. *Mater. Sci.* **2016**, *51*, 4996–5007. [CrossRef]

21. Chen, M.Y.; Gouné, M.; Verdier, M.; Bréchet, Y.; Yang, J.R. Interphase precipitation in vanadium-alloyed steels: Strengthening contribution and morphological variability with austenite to ferrite transformation. *Acta Mater.* **2014**, *64*, 78–92. [CrossRef]

22. Bu, F.Z.; Wang, X.M.; Yang, S.W.; Shang, C.J.; Misra, R.D.K. Contribution of interphase precipitation on yield strength in thermomechanically simulated Ti–Nb and Ti–Nb–Mo microalloyed steels. *Mater. Sci. Eng. A* **2015**, *620*, 22–29. [CrossRef]

23. Ariga, T.; Funakawa, Y.; Uchida, Y. High-Tensile-Strength Hot-Rolled Plated Steel Sheet and Method for Producing Same. Patent WO 2013/069210A1, 16 May 2013.

24. Xu, Y.; Zhang, W.; Sun, M.; Yi, H.; Liu, Z. The blocking effects of interphase precipitation on dislocations' movement in Ti-bearing micro-alloyed steels. *Mater. Lett.* **2015**, *139*, 177–181. [CrossRef]
25. Mukherjee, S.; Timokhina, I.B.; Zhu, C.; Ringer, S.P.; Hodgson, P.D. Three-dimensional atom probe microscopy study of interphase precipitation and nanocluster in thermomechanically treated titanium-molybdenum steels. *Acta Mater.* **2013**, *61*, 2521–2530. [CrossRef]
26. Hua, J.; Dua, L.X.; Wang, J.J. Effect of cooling procedure on microstructures and mechanical properties of hot rolled Nb-Ti bainitic high strength steel. *Mater. Sci. Eng.* **2012**, *554*, 79–85. [CrossRef]
27. Yen, H.W.; Chen, P.Y.; Huang, C.Y.; Yang, J.R. Interphase precipitation of nanometer-sized carbides in a titanium-molybdenum-bearing low-carbon steel. *Acta Mater.* **2011**, *59*, 6264–6274. [CrossRef]
28. Isasti, N.; Jorge-Badiola, D.; Taheri, M.L.; Uranga, P. Microstructural Features Controlling Mechanical Properties in Nb-Mo Microalloyed Steels. Part II: Impact Toughness. *Metall. Mater. Trans.* **2014**, *45*, 4972–4982. [CrossRef]
29. Gladman, T. *The Physical Metallurgy of Microalloyed Steels*, 2nd ed.; The Institute of Materials: London, UK, 1997.
30. Pickering, F.B. *Physical Metallurgy and the Design of Steels*; Applied Science Publishers Ltd.: London, UK, 1978.
31. Peng, Z.; Li, L.; Gao, J.; Huo, X. Precipitation strengthening of titanium microalloyed high-strength steel plates with isothermal treatment. *Mater. Sci. Eng.* **2016**, *657*, 413–421. [CrossRef]
32. Yakubtsov, I.A.; Boyd, J.D.; Liu, W.J.; Essadiqui, E. Strengthening mechanism in dual-phase acicular ferrite+M/A microstructure. In Proceedings of the 42nd Mechanical Working and Steel Processing Conference, Iron and Steel Society/AIME, Toronto, ON, Canada, 2000; pp. 429–439.
33. Pickering, F.B.; Gladman, T. *Metallurgical Developments in Carbon Steels*; Special Report No. 81; Iron and Steel Institute: London, UK, 1963.
34. Bush, M.E.; Kelly, P.M. Strengthening mechanisms in bainitic steels. *Acta Metall.* **1971**, *19*, 1363–1372. [CrossRef]
35. Gladman, T. Precipitation hardening in metals. *Mater. Sci. Technol.* **1999**, *15*, 30–36. [CrossRef]
36. Isasti, N.; Jorge-Badiola, D.; Taheri, M.L.; Uranga, P. Microstructural Features Controlling Mechanical Properties in Nb-Mo Microalloyed Steels. Part I: Yield Strength. *Metall. Mater. Trans.* **2014**, *45*, 4960–4971. [CrossRef]
37. Kim, Y.W.; Song, S.W.; Seo, S.J.; Hong, S.G.; Lee, C.S. Development of Ti and Mo micro-alloyed hot-rolled high strength sheet steel by controlling thermomechanical controlled processing schedule. *Mater. Sci. Eng.* **2013**, *565*, 430–438. [CrossRef]
38. Zhang, K.; Li, Z.; Wang, Z.; Sun, X.; Yong, Q. Precipitation behavior and mechanical properties of hot-rolled high strength Ti-Mo-bearing ferritic sheet steel: The great potential of nanometer-sized (Ti, Mo)C carbide. *Mater. Res.* **2016**, *31*, 1254–1263. [CrossRef]
39. Gutiérrez, I. Effect of microstructure on the impact toughness of Nb-microalloyed steel: Generalisation of existing relations from ferrite-pearlite to high strength microstructures. *Mater. Sci. Eng.* **2013**, *571*, 57–67. [CrossRef]
40. Gladman, T.; Holmes, B.; McIvor, I.D. *Effects of Second Phase Particles on Strength, Toughness and Ductility*; The Iron and Steel Institute: London, UK, 1971; p. 68.
41. Gray, J.M. Strength-Toughness Relations for Precipitation-Strengthened Low-Alloy Steels Containing Columbium. *Metall. Mater. Trans.* **1972**, *3*, 1495–1500. [CrossRef]
42. Pickering, F.B. The optimization of microstructures in steel and their relationship to mechanical properties. In *Hardenability Concepts with Applications to Steel*; Doane, D.V., Kirkaldy, J.S., Eds.; AIME: New York, NY, USA, 1978; pp. 179–228.

metals

Review

Transferring Nanoscale Bainite Concept to Lower C Contents: A Perspective

Carlos Garcia-Mateo [1,*], Georg Paul [2], Mahesh C. Somani [3], David A. Porter [3], Lieven Bracke [4], Andreas Latz [2], Carlos Garcia De Andres [1] and Francisca G. Caballero [1]

[1] Department of Physical Metallurgy, National Center for Metallurgical Research (CENIM-CSIC), Avenida Gregorio del Amo, 8, 28040 Madrid, Spain; cgda@cenim.csic.es (C.G.D.A.) fgc@cenim.csic.es (F.G.C.)
[2] Thyssenkrupp Steel Europe, Technology & Innovation, Modelling and Simulation, Kaiser-Wilhelm-Straße 100, 47166 Duisburg, Germany; georg.paul@thyssenkrupp.com (G.P.); andreas.latz@thyssenkrupp.com (A.L.)
[3] Materials Engineering and Production Technology, Faculty of Technology, University of Oulu, 90014 Oulu, Finland; mahesh.somani@oulu.fi (M.C.S.); david.porter@oulu.fi (D.A.P.)
[4] ArcelorMittal Global R&D Ghent, J.F. Kennedylaan, 9060 Zelzate, Belgium; lieven.bracke@arcelormittal.com
* Correspondence: cgm@cenim.csic.es; Tel.: +34-91-553-8900

Academic Editor: Håkan Hallberg
Received: 27 March 2017; Accepted: 28 April 2017; Published: 4 May 2017

Abstract: The major strengthening mechanisms in bainitic steels arise from the bainitic ferrite plate thickness rather than the length, which primarily determines the mean free slip distance. Both the strength of the austenite from where the bainite grows and the driving force of the transformation, are the two factors controlling the final scale of the bainitic microstructure. Usually, those two parameters can be tailored by means of selection of chemical composition and transformation temperature. However, there is also the possibility of introducing plastic deformation on austenite and prior to the bainitic transformation as a way to enhance both the austenite strength and the driving force for the transformation; the latter by introducing a mechanical component to the free energy change. This process, known as ausforming, has awoken a great deal of interest and it is the object of ongoing research with two clear aims. First, an acceleration of the sluggish bainitic transformation observed typically in high C steels (0.7–1 wt. %) transformed at relatively low temperatures. Second, to extend the concept of nanostructured bainite from those of high C steels to much lower C contents, 0.4–0.5 wt. %, keeping a wider range of applications in view.

Keywords: bainite; ausforming; kinetics; plate thickness

---

1. Structural Refinement of Bainitic Steels: General Considerations

Bainitic steels can be designed on the basis of the theory that predicts the highest temperature at which bainite (Bs) and martensite (Ms) can start to form in a steel of a given composition. These two temperatures constitute the upper and lower limits at which the isothermal heat treatment can be performed to generate bainite.

It has been reported that bainitic ferrite plate thickness depends primarily on three parameters, i.e., (1) the strength of the austenite at the transformation temperature, (2) the dislocation density in the austenite and (3) the chemical free energy change accompanying transformation [1–3]. In accord, a strong austenite possessing a high dislocation density and a large driving force results in finer plates. Austenite strength and dislocation density refine the structure by increasing the resistance to interface motion, and the thermodynamic driving force refines the structure by increasing the nucleation rate. All three factors—austenite strength, dislocation density and driving force—increase as the transformation temperature decreases, so a lower bainite transformation temperature leads to a reduction in the thickness of the bainitic ferrite plates.

2. Nanostructured Bainite

The concept exploited in a new generation of bainitic steels was the strengthening of the parent austenite by alloying, among others with high C levels, and lowering the transformation temperature as much as possible to give the austenite a higher strength. On this basis, high carbon (0.6–1 wt. %) high silicon (1.5–3 wt. %) steels were designed to produce incredibly fine plates of bainitic ferrite, 20–40 nm thick, separated by a percolating network of retained austenite after transformation at 200–350 °C [4,5]. The scale of this structure is such that it contains a remarkably large density of interfaces, making it very strong even without the presence of a substantial fraction of carbide precipitates. These steels present the highest strength/toughness combinations ever recorded in bainitic steels (2.5 GPa/30 MPa m$^{1/2}$) [6–8]. Recently, in the frame of a Research Fund for Coal and Steel (RFCS) project [9], these microstructures demonstrated superior potential for abrasive wear applications in large components, where a uniform microstructure free from residual stresses or without complex processing is required. Likewise, it was confirmed through industrial testing, that these new grades are on a par with significantly more expensive abrasive wear-resistant alloy steels [10].

However, there are two issues that limit the scope for their wider exploitation. First, as the same theory used for the design of such alloys predicts, the transformation slows down dramatically as the transformation temperature is reduced, and it may take as long as 10 to 20 h to fully transform to bainite [4,5]. Second, due to their high carbon content, the steels are difficult to weld because of the formation of untempered, brittle martensite in the coarse grained heat affected zones of the joints.

3. Transferring Nanostructured Bainite Concept

Therefore, the open question is how to transfer the nanoscale bainite concept to lower C contents, with enhanced transformation kinetics and weldability, in order to allow broader application, without deteriorating the superior combination of the mechanical and technological properties too much.

So far, different approaches have been tried with different levels of success.

3.1. Chemical Composition Modification

One approach was to introduce significant quantities of Mn and Ni, (2.3% Mn and 5% Ni wt. %) to try to lower the Bs–Ms temperature range and solid solution strengthen the austenite while maintaining a low C content (0.1–0.2 wt. %) [11]. It was found that the Bs temperature can indeed be suppressed in this way, but unlike the high carbon steels, the difference between Bs and Ms decreases drastically at high solute concentrations, see Figure 1. Furthermore, it has been suggested that the platelets of bainite tend to coalesce at low temperatures. This can be explained, theoretically, by an excess of available free energy during the transformation [12]. This coalescence counteracts the advantage of refining the bainite plates by their formation at low temperatures. The consequences of this coalescence on toughness are detrimental, but the influence on other properties requires further investigation [12–17].

In line with this approach, Soliman et al. [18] also reported the possibility of low temperature bainite in a 0.26 C wt. % steel heavily alloyed with Mn (3.44 wt. %) and Ni (1.85 wt. %) to suppress the transformation temperatures, Ms = 285 °C; the alloy also contained Co and Al with the purpose of accelerating the transformation [19]. Final reported bainitic ferrite plate thicknesses were of the order of 150 nm. The same authors performed a similar study [20] on higher C content steels (0.4–0.5 C wt. %) with Co and Al among other alloying elements, bringing down the Ms to 210 °C and producing bainitic ferrite plates between 40 and 120 nm. Qian et al. in their work [21] showed that in a 0.28C-1.96Mn-0.67Si-1.19Al-1.62Cr-0.34Ni-0.23Mo wt. % steel, it was possible to obtain bainite at temperatures around 320 °C, but there is no report of the scale of the microstructure.

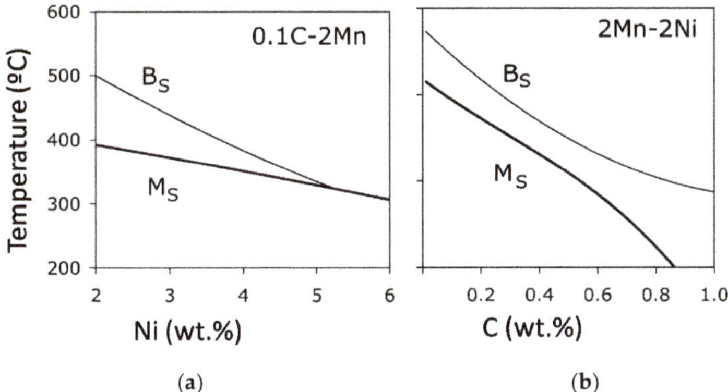

Figure 1. Thermodynamics calculation of bainite and martensite start transformation temperatures: (**a**) effect of Ni in Fe-0.1C-2Mn (wt. %) alloy system; (**b**) effect of C in Fe-2Mn-2Ni (wt. %) alloy system. Adapted from [11].

Wang et al. [22] and Long et al. [23] investigated an alloy (34MnSi-CrAlNiMo) containing 0.35C-1.5Si-1.5Mn-0.8Al-1.15Cr-0.4Mo all in wt. %. The measured *Ms* was found to be 310 °C and accordingly isothermal holding at 320–380 °C was applied to obtain bainite. The obtained microstructure was indeed a fine mixture of bainitic ferrites plates interwoven with thin films of retained austenite, see results of the measured bainitic ferrite plate and retained austenite film thickness, t_{BF} and t_{Ar} respectively, in Figure 2.

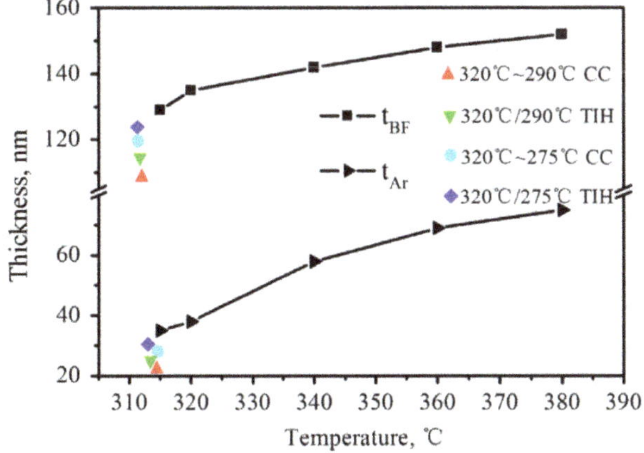

Figure 2. Relationship between t_{BF}, t_{Ar} and transformation temperature in 34MnSi-CrAlNiMo steel. TIH stands for Two Isothermal holding and CC for continuous cooling. Extracted from reference [23]. Reproduced with permission from Elsevier, 2017.

It is necessary to highlight that, in most of those works, the focus is put on attaining low temperature bainite, thus assuming that it will lead to nanostructured bainitic ferrite, which is not always the case. As already pointed out, low transformation temperatures will lead to a nanostructured bainite provided that the austenite from where bainite grows is sufficiently strong.

3.2. Heat Treatment Variations

There is another approach that aims at obtaining nanostructured bainite by the development of resourceful heat treatments. Wang et al. [24] by means of a multistep heat treatment, shown in Figure 3, managed to obtain an average plate thickness as low as 110 nm in a 0.30C-1.46Si-1.97Mn-1.50Ni-0.30Cr-0.96Cu-0.25Mo wt. % alloy. The concept lying beneath this multistep treatment is the continuous increase in the austenite C content due to partial bainitic transformation after each step, so that in the final steps, bainite will grow from an austenite whose C content is greater than that of the bulk, which also allows to carry on those final steps at temperatures that are lower than the *Ms* of the bulk alloy, but higher than the estimated *Ms* of the partially untransformed carbon-enriched austenite.

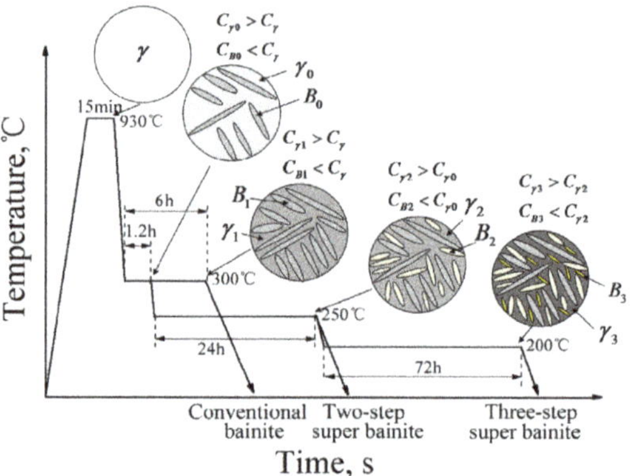

Figure 3. Heat treatment cycles and resultant microstructures during a multi-step low-temperature super-bainite transformation. Extracted from reference [24]. Reproduced with permission from Elsevier, 2017.

In line with this partial C enrichment of austenite to promote finer bainite, Li et al. [25] proposed a novel quenching and dynamic partitioning (Q–DP) process to obtain a fine microstructure comprising martensite-bainite laths and thin films of interlath retained austenite, in a medium C steel (0.3 C wt. %). The applied process is schematically shown in Figure 4.

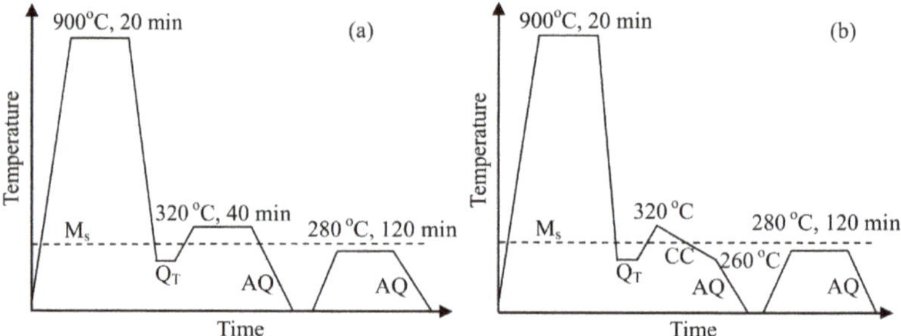

Figure 4. Different heat treatment processes used (**a**) Q&P process, (**b**) Q–DP process; QT: quenching temperature, CC: continuous cooling, AQ: air quenching. Extracted from reference [25]. Reproduced with permission from Elsevier, 2017.

Given that there exists the possibility of forming bainite by isothermal transformation at temperatures below the *Ms* [26–33], this is also becoming an attractive alternative, not only to accelerate the bainitic transformation, but also to obtain a finer microstructure in later stages of transformation. Figure 5 shows an example for 0.15 and 0.28 C steels transformed to bainite below the *Ms*; the author reports a decrease in the bainitic ferrite plate thickness of almost 40 nm in both cases, the final plate thickness being around 140–200 nm [26]. Bainite reaction occurs, after formation of some fraction of martensite, in a virtually identical manner as it would if that first transformed fraction had been obtained through isothermal bainite reaction. This is consistent with an autocatalytic effect that is identical whether one considers the influence of existing martensite or bainite laths on subsequent bainite lath nucleation [34].

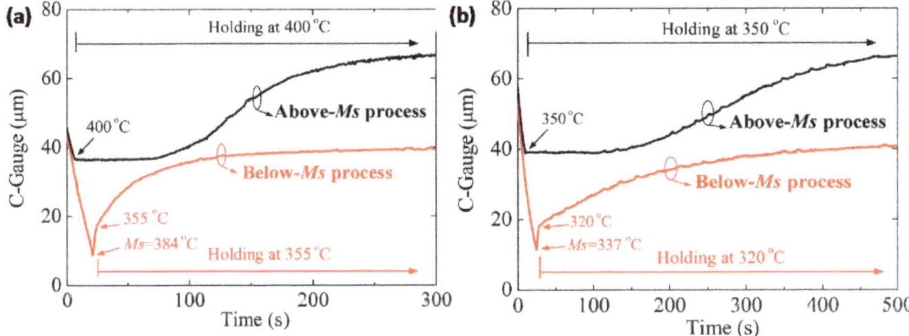

Figure 5. Dilation–time curves of 0.15 C wt. % steel (**a**), and 0.28 C wt. % steel (**b**) during cooling according to the above- and below-*Ms* austempering processes. Extracted from reference [26]. Reproduced with permission from Elsevier, 2017.

The reason why blocky austenite is undesirable is due to the fact that it has a lower chemical and mechanical stability and transforms to fresh, brittle high-carbon martensite absorbing low loads during toughness testing, thereby reducing the toughness. Bhadeshia and Edmonds [35] showed that desirable impact toughness was achieved in high-silicon bainitic steels containing films instead of blocks of retained austenite. Therefore, replacing austenitic blocks with a lath-like morphology is essential to improve steel toughness. Using multi-step isothermal bainite transformation in a medium carbon steel (0.30C-1.46Si-1.97Mn-1.50Ni-0.30Cr-0.96Cu-0.25Mo all in wt. %), Wang et al. [24] could almost eliminate the formation of blocky austenite through first partially transformed conventional low temperature bainite (300 °C/1.2 h), followed by the formation of higher volumes of nanoscale bainitic ferrite plates and retaining film-like austenite at still lower transformation temperatures (at 250/24 h and 200 °C/72 h), with a concomitant improvement in mechanical properties. A similar approach was adopted by Kim et al. [36]. On a 0.3C-1.5Si-1.5Mn wt. % steel and previously by Hase et al. [37] in developing nanostructured bainitic steel. They also reported an increase of the austenite content with thin film morphology and better ductility behaviour. It is even more important in case of coiled strips in industrial rolling, where microstructures comprising a range of bainitic lath thicknesses can form as cooling progresses. In yet another study on a 30MnSiCrAlNiMo 0.3C wt. % steel, Long et al. [38] showed that continuous cooling from *Ms* + 10 °C to *Ms* − 20 °C at the cooling rate of 0.5 °C/min produced fine plates of carbide free bainitic ferrite with thin films of retained austenite giving the best mechanical properties of any microstructures for the investigated steel.

3.3. Prior Austenite Grain Size Control (No Deformation)

Another set of approaches takes advantage of the fact that the *Ms* temperature decreases as the prior austenite grain size (PAGS) becomes smaller [39–42]. It is argued that this behavior is caused by grain boundary strengthening of the austenite, which makes martensite nucleation more difficult by providing a greater resistance to the motion of dislocations involved in the nucleation process comparable to the effect that solid solution strengthening has on the nucleation of martensite and bainite.

Conceptually, this idea can be summarized in the following sequence: the free energy change for the transformation of austenite to martensite must reach a critical value at the martensite start temperature (*Ms*), $\Delta G^{\gamma\alpha} < \Delta G_{Ms}$. Thus, when the PAGS decreases, there is a component of strengthening of the austenite that adds a mechanical component to the free energy change $\Delta G^{\gamma\alpha} + \Delta G_{MECH} < \Delta G_{Ms}$.

Figure 6 shows that the reduction of the transformation temperature is most pronounced when the PAGS < 10 µm [39,40].

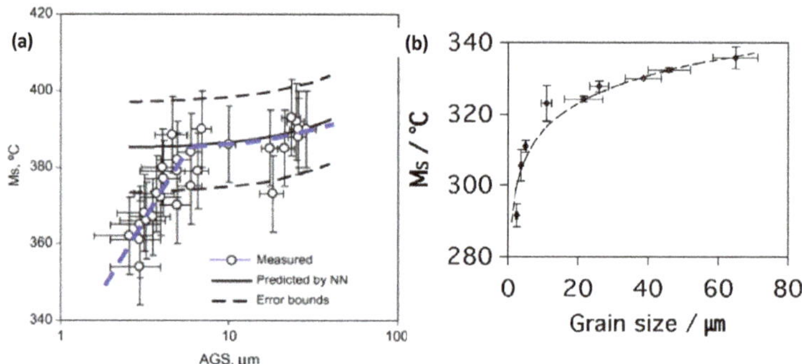

Figure 6. Influence of prior austenite grain size (PAGS) on the *Ms* temperature (**a**) comparison of experimental results and those obtained by a Neural Network model from reference [39]. Reproduced with permission from Elsevier, 2017. And (**b**) experimental work from reference [40]. Reproduced with permission from Elsevier, 2017.

In this sense, controlling and decreasing the austenitization temperature and time would affect the *Ms* temperature but only via the associated variation in austenite grain size. A very interesting side effect of the reduction of the PAGS is the associated increase in the bainite transformation kinetics, due to the increase in the number density of austenite grain surface nucleation sites. Consequently, there is a refinement of the size of the bainitic ferrite plates themselves, because an increase in the driving force stimulates a greater number density of plates [1,19].

3.4. Deformation of Austenite (AUSFORMING)

The option of applying a thermomechanical treatment to modify the PAGS also exists, where the combination of plastic deformation and precipitation on the nanoscale has been widely used in steels to control the prior austenite grain size at the micron level [43–45]. The most successful example is controlled rolling, with accelerated cooling for plates of low-C steels [43]. Although, in that particular case, the ultimate target of refinement is equiaxed ferrite, the principles remain the same. A decrease of the PAGS by controlled nucleation and growth of the recrystallized austenite and enhancement of the potent nucleation sites for ferrite, i.e., shear bands, dislocation substructures and stepped grain boundaries, has led to a minimum average ferrite grain size of 5 µm [43]. A combination of large-strain within the warm deformation regime and subsequent intercritical annealing, facilitate the formation of a dual phase microstructure with micron-sized ferrite [46].

From literature, it is clear that the studied thermomechanical routes can be divided into two groups: one, where the deformation is applied to austenite at high temperatures (above or below the recrystallization stop temperature and even in the warm deformation regime), and the other group, where deformation is applied to austenite at low temperatures, more precisely in the bay between the

C-curves of ferrite and bainite in the TTT (Time Temperature Transformation) diagram. Examples of both routes can be found in Figure 7.

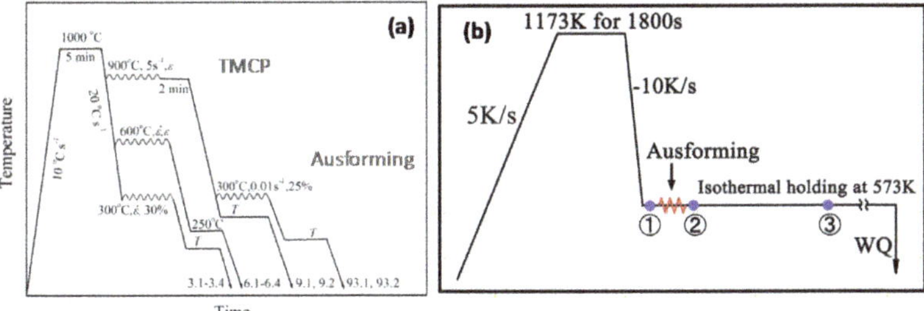

Figure 7. Examples of thermomechanical treatment applied to austenite prior to bainitic transformation. Extracted from references [47,48]. Reproduced with permission from Elsevier, 2017.

This same technique has received attention as a means of accelerating the sluggish bainitic transformation of (low temperature) nanostructured bainite and hence, different deformation routes have been tested, not only revealing accelerated transformation kinetics but also achieving an extra refinement of the microstructure [47–59]. While extensive work has been done on a single alloy containing 0.5 C wt. % [47,49,60–63], only in the case of Refs. [56,64], the carbon content was lowered to 0.2 wt. %. A summary of the microstructural characterization after ausforming at different temperatures prior to austempering at 355 °C (<*Ms* = 384 °C) in a 0.15 C steel is shown in Figure 8.

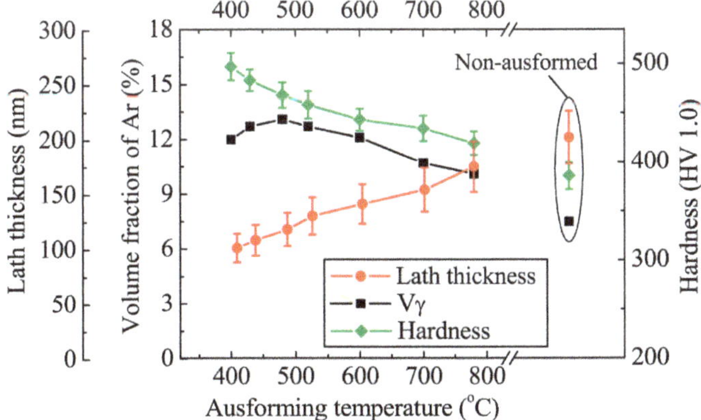

Figure 8. Average thickness of bainitic laths, volume fraction of retained austenite (Ar) and Vickers hardness as a function of ausforming temperature. The data for the non-ausformed sample are also shown for comparison. Extracted from reference [64]. Reproduced with permission from Elsevier, 2017.

An advantage of the low temperature process as compared with the higher temperature one is that the former can be employed to drastically reduce the number of bainite variants in a single austenite grain, leading to a reduction in the fraction of detrimental block-type retained austenite and an increase in the fraction of beneficial interlath films of retained austenite [48,50]. Gong et al. [48,50] studied the effects of ausforming temperature on bainite transformation and variant selection in high-carbon nanobainitic steel. Their results indicated that ausforming at a low temperature (300 °C) can accelerate bainite transformation and produce a strong variant selection, whereas ausforming at a high temperature (600 °C) has a weak influence.

Ausforming has also been shown to extend the temperature range available for isothermal transformation to bainite. Zhang et al. [47] showed that ausforming decreased the *Ms*, thereby enabling isothermal transformation at lower temperatures, thus facilitating the formation of nanostructured bainite. Besides this, decreasing the ausforming temperature and increasing the ausforming strain reduced the incubation time for bainite, refined the bainite lath size and enhanced the hardness. The decrease of *Ms* is explained by the mechanical part of the free energy change (ΔG_{MECH}).

3.5. Considerations

The decrease in *Ms* caused by ausforming and the consequent decrease in the transformation kinetics due to the decrease in the transformation temperature, would be compensated by the increased number of potent nucleation sites for bainite, grain boundary surface per volume unit and dislocations [19,47]. However, this same acceleration effect would also affect the reconstructive transformation to ferrite. Overall, it is therefore expected that ausforming will produce a general displacement of the CCT and TTT curves to shorter times provided that the dislocation density in the austenite is not too high.

The displacive transformations of austenite to martensite and bainite involve the coordinated movement of atoms across semicoherent glissile interfaces. Such movements cannot be sustained across grain boundaries, due to the loss of coherency, and therefore martensite and bainite plates are limited to single prior austenite grains. Isolated dislocations also hinder the progress of glissile interfaces, but they can often be accommodated by the transformation. However, it is well established [65,66] that if the strain in the austenite becomes sufficiently large, the motion of glissile interfaces becomes impossible, causing the transformation to halt. This applies to all cases involving the movement of glissile interfaces, whether their motion leads to a phase change [35,67] or simply to a reorientation of the lattice as occurs in mechanical twinning [68]. Therefore, sufficiently high dislocation densities in the austenite result in what is referred to as mechanical stabilization of the austenite [69–72], and it can lead to a complete cessation of transformation. Therefore, it is possible to define the critical strain for mechanical stabilization, as the strain necessary to apply to austenite so it is stabilized against bainitic or martensitic transformation [69]. Figure 9 shows an example of such calculations for a 316L stainless steel heavily alloyed with Ni, far away from the typical chemical compositions of bainitic steels but still representative of the existing problem with the mechanical stabilization of austenite.

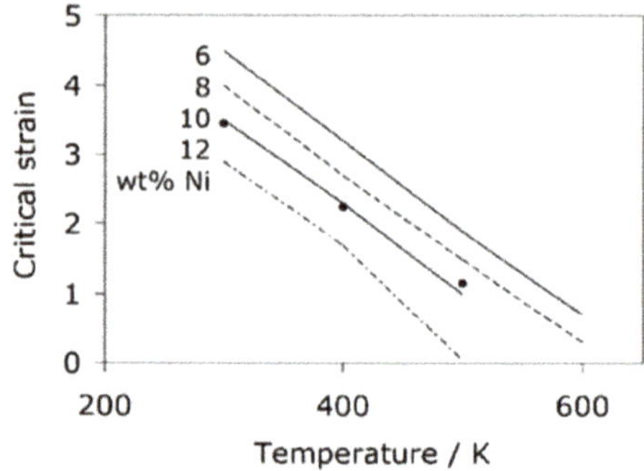

Figure 9. Critical values of plastic strain for mechanical stabilization of austenite in a 316L austenitic stainless steel. Extracted from reference [69]. Reproduced with permission from Taylor & Francis, 2017.

4. Conclusions

The nanostructured bainite concept invariably needs a strong austenite from where bainitic ferrite plates will grow. The advantages of transferring the nanoscale bainite concept to lower C (0.4–0.5 wt. %)

contents than those originally used for these types of alloys (1 wt. % C), would be to have alloys with enhanced transformation kinetics and weldability, allowing for much broader applications.

This work has revised different approaches that have pursued such an aim, including the most recent and novel, ausforming, where austenite is plastically deformed before bainitic transformation. Therefore, in accord with the literature review, to date, there does not exist a detailed study, where alloy design, process parameters, microstructure and mechanical response are investigated in order to obtain nanostructured bainite in a variety of medium C steels in conjunction with the ausforming process. Such a study is the subject of an ongoing project RFCS-2015-709607 that is expected to conclude with a recommendation of alloy chemical compositions and process parameters for industrial full-scale production.

Acknowledgments: The authors gratefully acknowledge the support of the European Research Fund for Coal and Steel for funding this research under the contracts RFCS-2015-709607.

Conflicts of Interest: The authors declare no conflict of interest.

References

1. Singh, S.B.; Bhadeshia, H.K.D.H. Estimation of bainite plate-thickness in low-alloy steels. *Mater. Sci. Eng. A* **1998**, *245*, 72–79. [CrossRef]
2. Cornide, J.; Garcia-Mateo, C.; Capdevila, C.; Caballero, F.G. An assessment of the contributing factors to the nanoscale structural refinement of advanced bainitic steels. *J. Alloys Compd.* **2013**, *577*, S43–S47. [CrossRef]
3. Caballero, F.G.; Garcia-Mateo, C.; Sourmail, T. Bainitic steel: Nanostructured. In *Encyclopedia of Iron, Steel, and Their Alloys*; Taylor & Francis Inc.: Abingdon, UK, 2016; pp. 271–290.
4. Garcia-Mateo, C.; Caballero, F.G.; Sourmail, T.; Smanio, V.; Garcia de Andres, C. Industrialised nanocrystalline bainitic steels. Design approach. *Int. J. Mater. Res.* **2014**, *105*, 725–734. [CrossRef]
5. Garcia-Mateo, C.; Sourmail, T.; Caballero, F.G.; Smanio, V.; Kuntz, M.; Ziegler, C.; Leiro, A.; Vuorinen, E.; Elvira, R.; Teeri, T. Nanostructured steel industrialisation: Plausible reality. *Mater. Sci. Technol.* **2014**, *30*, 1071–1078. [CrossRef]
6. Caballero, F.G.; Bhadeshia, H.K.D.H.; Mawella, K.J.A.; Jones, D.G.; Brown, P. Very strong low temperature bainite. *Mater. Sci. Technol.* **2002**, *18*, 279–284. [CrossRef]
7. Garcia-Mateo, C.; Caballero, F.G. Ultra-high-strength bainitic steels. *ISIJ Int.* **2005**, *45*, 1736–1740. [CrossRef]
8. Garcia-Mateo, C.; Caballero, F.G.; Sourmail, T.; Kuntz, M.; Cornide, J.; Smanio, V.; Elvira, R. Tensile behaviour of a nanocrystalline bainitic steel containing 3 wt % silicon. *Mater. Sci. Eng. A* **2012**, *549*, 185–192. [CrossRef]
9. Sourmail, T.; Smanio, V.; Ziegler, C.; Heuer, V.; Kuntz, M.; Caballero, F.G.; Garcia-Mateo, C.; Cornide, J.; Elvira, R.; Leiro, A.; et al. *Novel Nanostructured Bainitic Steel Grades to Answer the Need for High-Performance Steel Components (Nanobain)*; Rfsr-ct-2008-00022; European Commission: Luxembourg, 2013; p. 129.
10. Leiro, A.; Vuorinen, E.; Sundin, K.G.; Prakash, B.; Sourmail, T.; Smanio, V.; Caballero, F.G.; Garcia-Mateo, C.; Elvira, R. Wear of nano-structured carbide-free bainitic steels under dry rolling-sliding conditions. *Wear* **2013**, *298*, 42–47. [CrossRef]
11. Yang, H.S.; Bhadeshia, H.K.D.H. Designing low carbon, low temperature bainite. *Mater. Sci. Technol.* **2008**, *24*, 335–342. [CrossRef]
12. Ohmori, Y.; Ohtsubo, H.; Jung, Y.C.; Okaguchi, S.; Ohtani, H. Morphology of bainite and widmanstätten ferrite. *Metall. Mater. Trans. A* **1994**, *25*, 1981–1989. [CrossRef]
13. Pak, J.; Suh, D.W.; Bhadeshia, H.K.D.H. Promoting the coalescence of bainite platelets. *Scr. Mater.* **2012**, *66*, 951–953. [CrossRef]
14. Bhadeshia, H.K.D.H.; Keehan, E.; Karlsson, L.; Andrén, H.O. Coalesced bainite. *Trans. Indian Inst. Met.* **2006**, *59*, 689–694.
15. Keehan, E.; Karlsson, L.; Andren, H.O. Influence of carbon, manganese and nickel on microstructure and properties of strong steel weld metals: Part 1—Effect of nickel content. *Sci. Technol. Weld. Join.* **2006**, *11*, 1–8. [CrossRef]
16. Pak, J.H.; Bhadeshia, H.K.D.H.; Karlsson, L.; Keehan, E. Coalesced bainite by isothermal transformation of reheated weld metal. *Sci. Technol. Weld. Join.* **2008**, *13*, 593–597. [CrossRef]

17. Caballero, F.G.; Chao, J.; Cornide, J.; Garcia-Mateo, C.; Santofimia, M.J.; Capdevila, C. Toughness deterioration in advanced high strength bainitic steels. *Mater. Sci. Eng. A* **2009**, *525*, 87–95. [CrossRef]

18. Soliman, M.; Mostafa, H.; El-Sabbagh, A.S.; Palkowski, H. Low temperature bainite in steel with 0.26 wt % C. *Mater. Sci. Eng. A* **2010**, *527*, 7706–7713. [CrossRef]

19. Garcia-Mateo, C.; Caballero, F.; Bhadeshia, H. Acceleration of low-temperature bainite. *ISIJ Int.* **2003**, *43*, 1821–1825. [CrossRef]

20. Soliman, M.; Palkowski, H. Ultra-fine bainite structure in hypo-eutectoid steels. *ISIJ Int.* **2007**, *47*, 1703–1710. [CrossRef]

21. Qian, L.; Zhou, Q.; Zhang, F.; Meng, J.; Zhang, M.; Tian, Y. Microstructure and mechanical properties of a low carbon carbide-free bainitic steel co-alloyed with Al and Si. *Mater. Des.* **2012**, *39*, 264–268. [CrossRef]

22. Wang, Y.H.; Zhang, F.C.; Wang, T.S. A novel bainitic steel comparable to maraging steel in mechanical properties. *Scr. Mater.* **2013**, *68*, 763–766. [CrossRef]

23. Long, X.Y.; Kang, J.; Lv, B.; Zhang, F.C. Carbide-free bainite in medium carbon steel. *Mater. Des.* **2014**, *64*, 237–245. [CrossRef]

24. Wang, X.L.; Wu, K.M.; Hu, F.; Yu, L.; Wan, X.L. Multi-step isothermal bainitic transformation in medium-carbon steel. *Scr. Mater.* **2014**, *74*, 56–59. [CrossRef]

25. Li, Q.; Huang, X.; Huang, W. Microstructure and mechanical properties of a medium-carbon bainitic steel by a novel quenching and dynamic partitioning (Q-DP) process. *Mater. Sci. Eng. A* **2016**, *662*, 129–135. [CrossRef]

26. Zhao, L.; Qian, L.; Meng, J.; Zhou, Q.; Zhang, F. Below-Ms austempering to obtain refined bainitic structure and enhanced mechanical properties in low-C high-Si/Al steels. *Scr. Mater.* **2016**, *112*, 96–100. [CrossRef]

27. Da Silva, E.P.; De Knijf, D.; Xu, W.; Föjer, C.; Houbaert, Y.; Sietsma, J.; Petrov, R. Isothermal transformations in advanced high strength steels below martensite start temperature. *Mater. Sci. Technol.* **2015**, *31*, 808–816. [CrossRef]

28. Hofer, C.; Leitner, H.; Winkelhofer, F.; Clemens, H.; Primig, S. Structural characterization of "carbide-free" bainite in a Fe–0.2C–1.5Si–2.5Mn steel. *Mater. Charact.* **2015**, *102*, 85–91. [CrossRef]

29. Yoshikawa, N.; Kobayashi, J.; Sugimoto, K.-I. Notch-fatigue properties of advanced trip-aided bainitic ferrite steels. *Metall. Mater. Trans. A* **2012**, *43*, 4129–4136. [CrossRef]

30. Sugimoto, K.I. Fracture strength and toughness of ultra high strength trip aided steels. *Mater. Sci. Technol.* **2009**, *25*, 1108–1117. [CrossRef]

31. Kolmskog, P.; Borgenstam, A.; Hillert, M.; Hedström, P.; Babu, S.S.; Terasaki, H.; Komizo, Y.I. Direct observation that bainite can grow below Ms. *Metall. Mater. Trans. A* **2012**, *43*, 4984–4988. [CrossRef]

32. van Bohemen, S.M.C.; Santofimia, M.J.; Sietsma, J. Experimental evidence for bainite formation below Ms in Fe-0.66C. *Scr. Mater.* **2008**, *58*, 488–491. [CrossRef]

33. Samanta, S.; Biswas, P.; Giri, S.; Singh, S.B.; Kundu, S. Formation of bainite below the Ms temperature: Kinetics and crystallography. *Acta Mater.* **2016**, *105*, 390–403. [CrossRef]

34. Smanio, V.; Sourmail, T. Effect of partial martensite transformation on bainite reaction kinetics in different 1%C steels. *Solid State Phenom.* **2011**, *172–174*, 821–826. [CrossRef]

35. Bhadeshia, H.K.D.H.; Edmonds, D.V. Bainite in silicon steels: New composition-property approach. Part 2. *Met. Sci.* **1983**, *17*, 420–425. [CrossRef]

36. Kim, K.W.; Il Kim, K.; Lee, C.H.; Kang, J.Y.; Lee, T.H.; Cho, K.M.; Oh, K.H. Control of retained austenite morphology through double bainitic transformation. *Mater. Sci. Eng. A* **2016**, *673*, 557–561. [CrossRef]

37. Hase, K.; Garcia-Mateo, C.; Bhadeshia, H.K.D.H. Bimodal size-distribution of bainite plates. *Mater. Sci. Eng. A* **2006**, *438*, 145–148. [CrossRef]

38. Long, X.Y.; Zhang, F.C.; Kang, J.; Lv, B.; Shi, X.B. Low-temperature bainite in low-carbon steel. *Mater. Sci. Eng. A* **2014**, *594*, 344–351. [CrossRef]

39. García-Junceda, A.; Capdevila, C.; Caballero, F.G.; de Andrés, C.G. Dependence of martensite start temperature on fine austenite grain size. *Scr. Mater.* **2008**, *58*, 134–137. [CrossRef]

40. Yang, H.-S.; Bhadeshia, H.K.D.H. Austenite grain size and the martensite-start temperature. *Scr. Mater.* **2009**, *60*, 493–495. [CrossRef]

41. Lee, S.-J.; Park, K.-S. Prediction of martensite start temperature in alloy steels with different grain sizes. *Metall. Mater. Trans. A* **2013**, *44*, 3423–3427. [CrossRef]

42. Yang, H.S.; Suh, D.W.; Bhadeshia, H.K.D.H. More complete theory for the calculation of the martensite-start temperature in steels. *ISIJ Int.* **2012**, *52*, 164–166. [CrossRef]

43. Maki, T. Current state and future prospect of microstructure control in steels. *Tetsu To Hagane-J. ISIJ* **1995**, *81*, N547–N555. [CrossRef]

44. Krauss, G. *Steels: Heat Treatment and Processing Principles*; ASM International: Materials Park, OH, USA, 1989.

45. Calcagnotto, M.; Ponge, D.; Raabe, D. Effect of grain refinement to 1 μm on strength and toughness of dual-phase steels. *Mater. Sci. Eng. A* **2010**, *527*, 7832–7840. [CrossRef]

46. Calcagnotto, M.; Adachi, Y.; Ponge, D.; Raabe, D. Deformation and fracture mechanisms in fine- and ultrafine-grained ferrite/martensite dual-phase steels and the effect of aging. *Acta Mater.* **2011**, *59*, 658–670. [CrossRef]

47. Zhang, M.; Wang, Y.H.; Zheng, C.L.; Zhang, F.C.; Wang, T.S. Effects of ausforming on isothermal bainite transformation behaviour and microstructural refinement in medium-carbon Si-Al-rich alloy steel. *Mater. Des.* **2014**, *62*, 168–174. [CrossRef]

48. Gong, W.; Tomota, Y.; Adachi, Y.; Paradowska, A.M.; Kelleher, J.F.; Zhang, S.Y. Effects of ausforming temperature on bainite transformation, microstructure and variant selection in nanobainite steel. *Acta Mater.* **2013**, *61*, 4142–4154. [CrossRef]

49. Zhang, M.; Wang, T.S.; Wang, Y.H.; Yang, J.; Zhang, F.C. Preparation of nanostructured bainite in medium-carbon alloy steel. *Mater. Sci. Eng. A* **2013**, *568*, 123–126. [CrossRef]

50. Gong, W.; Tomota, Y.; Koo, M.S.; Adachi, Y. Effect of ausforming on nanobainite steel. *Scr. Mater.* **2010**, *63*, 819–822. [CrossRef]

51. Chakraborty, J.; Bhattacharjee, D.; Manna, I. Development of ultrafine bainite + martensite duplex microstructure in sae 52100 bearing steel by prior cold deformation. *Scr. Mater.* **2009**, *61*, 604–607. [CrossRef]

52. Chakraborty, J.; Manna, I. Development of ultrafine ferritic sheaves/plates in sae 52100 steel for enhancement of strength by controlled thermomechanical processing. *Mater. Sci. Eng. A* **2012**, *548*, 33–42. [CrossRef]

53. Lonardelli, I.; Bortolotti, M.; Van Beek, W.; Girardini, L.; Zadra, M.; Bhadeshia, H.K.D.H. Powder metallurgical nanostructured medium carbon bainitic steel: Kinetics, structure, and in situ thermal stability studies. *Mater. Sci. Eng. A* **2012**, *555*, 139–147. [CrossRef]

54. Zhang, M.; Wang, Y.H.; Zheng, C.L.; Zhang, F.C.; Wang, T.S. Austenite deformation behavior and the effect of ausforming process on martensite starting temperature and ausformed martensite microstructure in medium-carbon Si-Al-rich alloy steel. *Mater. Sci. Eng. A* **2014**, *596*, 9–14. [CrossRef]

55. He, J.; Zhao, A.; Huang, Y.; Zhi, C.; Zhao, F. Acceleration of Bainite Transformation at Low Temperature by Warm Rolling Process. *Mater. Today Proc.* **2015**, *2*, S289–S294. [CrossRef]

56. Timokhina, I.; Beladi, H.; Xiong, X.; Hodgson, P. Effect of composition and processing parameters on the formation of nano-bainite in advanced high strength steels. *J. Iron Steel Res. Int.* **2011**, *18*, 238–241.

57. Golchin, S.; Avishan, B.; Yazdani, S. Effect of 10% ausforming on impact toughness of nano bainite austempered at 300 °C. *Mater. Sci. Eng. A* **2016**, *656*, 94–101. [CrossRef]

58. He, J.; Zhao, A.; Zhi, C.; Fan, H. Acceleration of nanobainite transformation by multi-step ausforming process. *Scr. Mater.* **2015**, *107*, 71–74. [CrossRef]

59. Kabirmohammadi, M.; Avishan, B.; Yazdani, S. Transformation kinetics and microstructural features in low-temperature bainite after ausforming process. *Mater. Chem. Phys.* **2016**, *184*, 306–317. [CrossRef]

60. Zhou, M.; Xu, G.; Wang, L.; Hu, H. Combined effect of the prior deformation and applied stress on the bainite transformation. *Met. Mater. Int.* **2016**, *22*, 956–961. [CrossRef]

61. Zhi, C.; Zhao, A.; He, J.; Yang, H.; Qi, L. Effects of the multi-step ausforming process on the microstructure evolution of nanobainite steel. In Proceedings of the International Conference on Advanced Materials, Structures and Mechanical Engineering, Incheon, Korea, 29–31 May 2015; Kaloop, M., Ed.; CRC Press: Boca Raton, FL, USA, 2016; pp. 399–403.

62. Zhou, M.; Xu, G.; Zhang, Y.; Xue, Z. The effects of external compressive stress on the kinetics of low temperature bainitic transformation and microstructure in a superbainite steel. *Int. J. Mater. Res.* **2015**, *106*, 1040–1045. [CrossRef]

63. Hu, H.; Xu, G.; Wang, L.; Zhou, M. Effects of strain and deformation temperature on bainitic transformation in a Fe-C-Mn-Si alloy. *Steel Res. Int.* **2016**, *88*. [CrossRef]

64. Zhao, L.; Qian, L.; Liu, S.; Zhou, Q.; Meng, J.; Zheng, C.; Zhang, F. Producing superfine low-carbon bainitic structure through a new combined thermo-mechanical process. *J. Alloys Compd.* **2016**, *685*, 300–303. [CrossRef]

65. Tsuzaki, K.; Fukasaku, S.-I.; Tomota, Y.; Maki, T. Effect of prior deformation of austenite on the $\gamma \rightarrow \varepsilon$ martensitic transformation in Fe-Mn alloys. *Mater. Trans. JIM* **1991**, *32*, 222–228. [CrossRef]

66. Bhadeshia, H.K.D.H. The bainite transformation: Unresolved issues. *Mater. Sci. Eng. A* **1999**, *273–275*, 58–66. [CrossRef]

67. Shipway, P.H.; Bhadeshia, H.K.D.H. The mechanical stabilisation of widmanstatten ferrite. *Mater. Sci. Eng. A* **1997**, *223*, 179–185. [CrossRef]

68. Christian, J.W.; Mahajan, S. Deformation twinning. *Prog. Mater. Sci.* **1995**, *39*, 1–157. [CrossRef]

69. Chatterjee, S.; Wang, H.S.; Yang, J.R.; Bhadeshia, H.K.D.H. Mechanical stabilisation of austenite. *Mater. Sci. Technol.* **2006**, *22*, 641–644. [CrossRef]

70. Shipway, P.H.; Bhadeshia, H.K.D.H. Mechanical stabilisation of bainite. *Mater. Sci. Technol.* **1995**, *11*, 1116–1128. [CrossRef]

71. Maalekian, M.; Kozeschnik, E.; Chatterjee, S.; Bhadeshia, H.K.D.H. Mechanical stabilisation of eutectoid steel. *Mater. Sci. Technol.* **2007**, *23*, 610–612. [CrossRef]

72. Yi, H.L.; Lee, K.Y.; Bhadeshia, H.K.D.H. Mechanical stabilisation of retained austenite in δ-trip steel. *Mater. Sci. Eng. A* **2011**, *528*, 5900–5903. [CrossRef]

Article

The Nitrocarburising Response of Low Temperature Bainite Steel

Daniel Fabijanic, Ilana Timokhina *, Hossein Beladi and Peter Hodgson

Institute for Frontier Materials, Deakin University; Geelong 3220, Australia;
daniel.fabijanic@deakin.edu.au (D.F.); hossein.beladi@deakin.edu.au (H.B.);
peter.hodgson@deakin.edu.au (P.H.)
* Correspondence: ilana.timokhina@deakin.edu.au; Tel.: +61-3-52272413

Received: 31 May 2017; Accepted: 21 June 2017; Published: 26 June 2017

Abstract: The nitrocarburising response of low transformation temperature ultrafine and nanoscale bainitic steel was investigated and compared with martensite and pearlite from the same steel composition. It was found that the retained austenite content of the bainitic steel dictated the core hardness after nitrocarburising. The refined bainitic structure showed improvements in the nitriding depth and hardness of the nitrocarburised layer, compared to coarser grained martensitic and pearlitic structures, possibly due to the fine structure and the distribution of nitride forming elements.

Keywords: low temperature bainite; nitrocarburising; surface modification; retained austenite; bainitic ferrite; transmission electron microscopy

---

1. Introduction

Recent developments in the compositional design and thermal treatment of high-strength steels have challenged the traditional compromise between strength and toughness. By refining bainitic ferrite laths to a nanometer scale, a new generation of nanostructured bainite steels have emerged, offering a highly desirable combination of properties: high strength (2.3 GPa), toughness (30 MPa m$^{1/2}$) and ductility (30%) in both quasistatic and dynamic loading conditions [1–3]. The potential application range of these steels is broad, encompassing the transport, construction and defense industries. In all of these applications, there is potential for improved performance by enhancing the surface properties. There have been studies of microstructure and property relationships [1–4], crystallography [5,6], wear behavior [7,8], and thermal tempering [9,10] of nano-bainitic steels (henceforth termed low temperature bainite) but it has been no attempt to enhance its surface properties.

Ferritic nitrocarburising is a thermo-chemical surface treatment designed to diffuse nitrogen and carbon into the surface of ferrous materials [11]. A surface structure typically consists of a hard outer iron carbonitride (ε-Fe$_{2-3}$(C, N)$_{1-x}$) compound layer and a tough inner nitrogen-enriched diffusion zone [11]. The addition of a carbon-bearing gas to the treatment atmosphere distinguishes nitrocarburising from nitriding, and increases the nitrogen diffusion kinetics. Due to the compatibility between the kinetics of compound layer growth and the tempering behavior of the low temperature bainitic steel, nitrocarburising was chosen in preference to nitriding in this study. The growth of both the compound layer and diffusion zone depends on the process conditions (time, temperature, nitriding potential) and nitrogen diffusivity. The latter factor is strongly influenced by the grain structure of the materials. It has been found [12–15] that materials that have a large number of grain boundaries (e.g., an ultrafine or nanocrystalline material) exhibit faster nitriding kinetics, where the grain boundaries act as diffusion channels. Moreover, low temperature bainitic steel exhibits excellent tempering resistance [16], which is beneficial for the nitrocarburasing treatment.

This study explores the nitrocarburising response of the low temperature bainitic steel and compares this response to coarser structures (martensite and pearlite) of the same material composition.

2. Materials and Methods

The steel composition used in this study (0.79C-1.5Si-1.98Mn-0.98Cr-0.24Mo-1.06Al-1.58Co wt %) was based on a composition proposed in [1]. To form low temperature bainite, the steel was austenitized at 1100 °C for 30 min, followed immediately by an isothermal heat treatment between the M_s (155 °C) and B_s (385 °C) temperatures. The starting bainite (B_s) and martensite (M_s) phase transformation temperatures were chosen to be 385 and 155 °C, respectively, based on previously published work [17]. Two isothermal treatments were used; 200 °C for 10 days ('200' bainite) and 350 °C for 1 day ('350' bainite) [6]. The air and furnace cooling from the austenitizing temperature were used to form martensite ('M') and coarse pearlite ('P'), respectively.

Tempering of the bainitic steels was performed in an air furnace for various times (10–7200 min) and temperatures (250–600 °C), and the resulting core hardness determined by Vickers indention using a 20 kg load and a 10 s dwell time. The tempering response of the bainitic steels was used to determine the nitrocarburising parameters, as will be discussed later. Samples were surface polished (1200 grade SiC), ultrasonically cleaned in ethanol and nitrocarburised in a fluid bed reactor at 525 °C for 4 h in a gaseous atmosphere of 45% ammonia, 3% CO_2 and balance nitrogen. The microstructure of the core and nitrocarburised region was observed by light optical microscopy, and detailed microstructural features were determined by Transmission Electron Microscopy (TEM).

TEM foils, discs 3 mm in diameter, were mechanically ground to a thickness of ~0.07 mm and then twin-jet electropolished using a solution containing 5% perchloric acid and 95% methanol at a temperature of -25 °C and a voltage of 50 V. TEM examination of thin foils was performed using a Philips CM 20 microscope (Philips, Electronic Instruments Corporation, Mahwah, NJ, USA) operated at 200 kV.

The retained austenite volume fractions in the core material after nitrocarburising treatment were determined using a Phillips PW 1130 diffractometer with graphite monochromated CuK_α radiation at 40 kV and 25 mA in the 2θ range of 40–90° at a rate of 0.5°/min and a step size of 0.05° (Philips, Electronic Instruments Corporation, Mahwah, NJ, USA). A Sietronics X-Ray Diffraction (XRD) automation system was used for data collection, with analysis by XRD Traces V. 5.2 software (5.2, GBC Scientific Equipment, Hampshire, IL, USA). The retained austenite content was calculated from the integrated intensities of the $(200)_\alpha$, $(211)_\alpha$, $(200)_\gamma$ and $(220)_\gamma$ peaks using the direct comparison method [18]. Micro-Vickers hardness profiles of the nitrocarburised surfaces were performed on a mounted cross-section using a 50 gf load and a 10 s dwell time. A quantitative depth profile for nitrogen content was obtained using a Leco GDS850 glow discharge optical emission spectrometer (LECO Corporation, Saint Joseph, MI, USA).

3. Results and Discussion

3.1. Characterization of Initial Micostructures

A detailed study of the microstructural characteristics and crystallographic analysis of the low temperature bainite formed at 200 and 350 °C has been reported in [4–6]. Isothermal treatment at 200 °C produced a fine lamella microstructure of alternating films of austenite (21 ± 2% vol. fraction, 30 ± 5 nm thick) and bainitic ferrite (60 ± 10 nm thick). The bainitic ferrite and retained austenite layers formed colonies or sheaves, where the layers or sub-units consisting of bainitic ferrite and residual austenite are parallel to each other (Figure 1a).

Increasing the isothermal treatment to 350 °C resulted in formation of bainite packets consisting of nanolayers of bainitic ferrite with thickness from 200 to 400 nm, and retained austenite layers with an average thickness of 70 ± 30 nm (Figure 1b). The bainite/retained austenite sub-unit formed at this temperature was coarser than the similar sub-unit after isothermal holding at 200 °C. The volume fraction of austenite was 53 ± 1%.

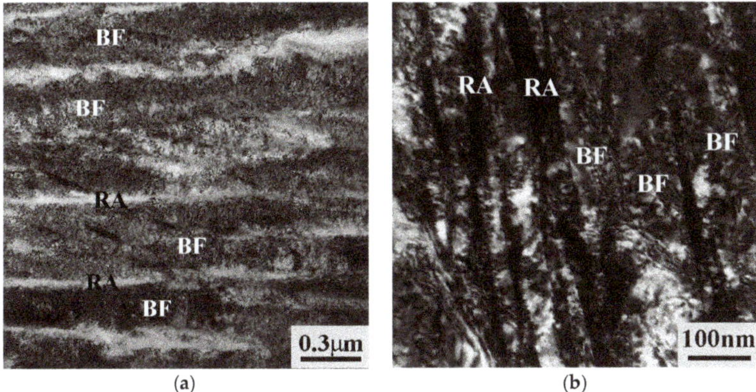

Figure 1. TEM (Transmission Electron Microscopy) micrographs of the microstructure formed after 200 °C (**a**) and 350 °C (**b**) isothermal treatment. RA is retained austenite and BF is bainitic ferrite.

The microstructure of the air-cooled specimen contained both plate and lath martensite morphology in approximately equal proportions (Figure 2a,b). Plate martensite was previously described as plates having a thickness of several microns and a length of tens of microns without further structural subdivisions of these plates (Figure 2a) [19]. In contrast, lath martensite forms the same three-level hierarchy as the bainitic steels, ultimately forming laths of martensite with sub-micrometer thickness (Figure 2b) [20]. The furnace cooled structure was coarse pearlite. The core hardness of each microstructural variant was 380 ± 6 ('350'), 675 ± 7 ('200'), 365 ± 7 ('P') and 654 ± 5 ('M') HV_{50gf}.

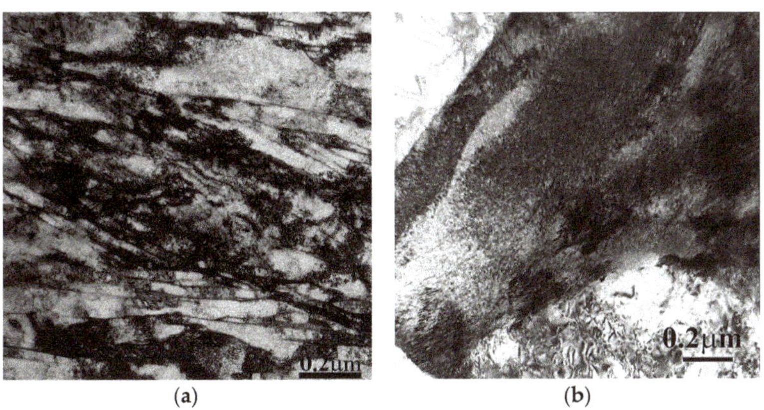

Figure 2. TEM images of plate (**a**) and lath (**b**) martensite.

3.2. Effect of Tempering on Microstructure and Properties

The tempering behavior is very different for the two bainitic steel variants (Figure 3). A reduction in hardness with time and temperature occurred during the tempering of '200' bainite. In contrast, the '350' bainite exhibited a hardening peak, which occurred earlier with increased tempering temperature. A nitrocarburising process window is indicated in Figure 3. This window provided a balance between favorable nitrocarburising parameters and retaining core hardness.

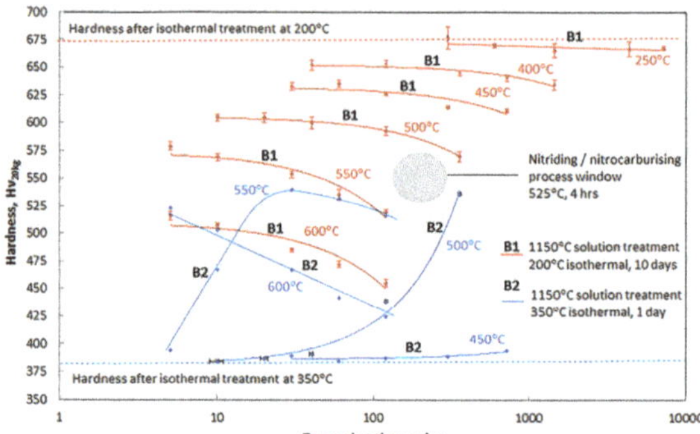

Figure 3. The hardness response of the '200' (B1) and '350' (B2) bainitic steels during isothermal tempering at various times and temperatures. A nitrocarburising time-temperature process window is indicated.

The retained austenite content of the '200' bainite core reduced from $21 \pm 2\%$ to $12 \pm 1\%$ after nitrocarburising at 525 °C for 4 h. TEM imaging revealed several microstructural features: (i) the lamellar structure was largely retained, (ii) spherical cementite particles (10–50 nm diameters) formed within the bainitic ferrite (Figure 4a,b), (iii) the dislocation density of the bainitic ferrite decreased, and (iv) some of the retained austenite films decomposed with formation of cementite (Figure 4a,b). Our recent study of the initial microstructure of similar steels under similar isothermal conditions using TEM and Atom Probe Tomography showed [4] a high volume fraction of clusters and carbides in bainitic ferrite, and low stability of the retained austenite. This was associated with high dislocation density in bainitic ferrite and preferential segregation of carbon to the dislocations in ferrite rather than diffusing to the remaining austenite [4]. Tempering of this microstructure promoted the fast carbon diffusion that led to the formation of spherical carbides in bainitic ferrite and decomposition of the retained austenite due to supersaturation of the retained austenite with carbon. Hence, the general decrease in hardness with time and temperature for the '200' bainite is reasonable.

In contrast, the '350' bainite experienced significant core hardening of nearly 200 HV after nitrocarburising at 525 °C for 4 h (Figure 3). Certainly, the hardness increase is related to the large decrease in retained austenite from $53 \pm 1\%$ to $4 \pm 1\%$. The austenite decomposed to form carbides of lenticular morphology with a width of ~10 nm (Figure 4c). TEM also confirmed the formation of spherical carbides in the bainitic ferrite (Figure 4d). Other studies [9,10] have shown that blocky austenite in nanostructured bainite can form pearlite with an extremely fine lamellar spacing when tempered above 500 °C. It would appear that the thicker initial austenite films in the '350' bainite decompose in a similar manner to blocky austenite. The conversion of half of the structure from austenite to extremely fine pearlite during nitrocarburising could account for the hardening. Another possibility is that the austenite has decomposed to a mixture of ferrite and lenticular carbides. The nitrocarburising could result in an increase in the M_s temperature in the retained austenite films, leading to martensite transformation. This would give a large increase in hardness. At this stage, though, it is not clear exactly what mechanism is responsible, but the important issue here is that it is possible to obtain a high core strength in the '350' bainite.

The decomposition of the retained austenite with an increase in tempering temperature has been reported in [9,10]. However, it is still unclear how the ferrite component can accommodate the extremely high quantity of carbon during retained austenite decomposition.

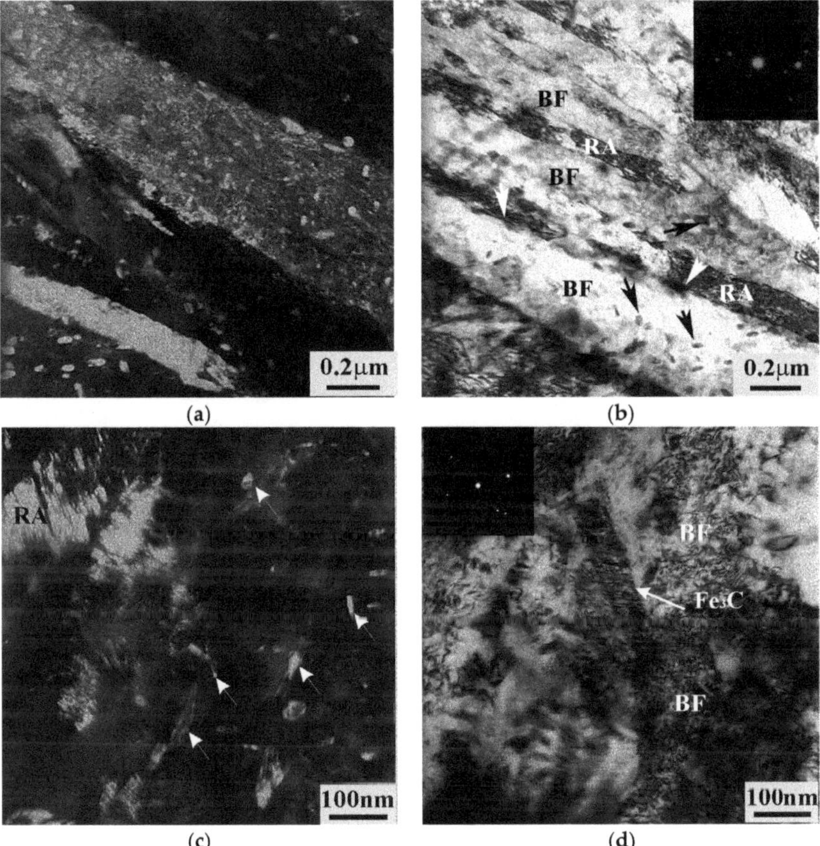

Figure 4. Bright (**a,c**) and dark (**b,d**) TEM images of the bainite core microstructure formed after nitrocarburising at 525 °C for 4 h in '200' (**a,b**) and '350' (**c,d**) bainite: (**a,b**) formation of spherical carbides in bainitic ferrite (dark arrows) and decomposition of retained austenite (white arrows), (**c**) decomposition of the retained austenite with formation of lenticular carbides (white arrow) and (**d**) spherical carbides in the bainitic ferrite.

After nitrocarburising the core hardness was highest in the bainitic steels at 578 ± 8 and 542 ± 4 HV$_{50gf}$ for the 350 and 200 °C isothermally treated variants, respectively (Figure 3). The core hardness of the martensitic steel was 514 ± 4 HV$_{50gf}$ and the core hardness of the pearlitic steel was 369 ± 9 HV$_{50gf}$. This demonstrates the good tempering resistance of the nanobainitic steels, a property that Cabellero [10] has attributed to the complex partitioning and microstructural changes that occur during reheating. High core hardness is desirable for a nitrocarburised component, as it provides good core strength and load carrying capacity. If core hardness and strength are desired, then low temperature bainite is an appropriate starting material. From a processing standpoint, it is significant that the '350' variant produced the higher core hardness, as the treatment time is shorter by a factor of 10. However, other mechanical properties are also significant, notably ductility and toughness, which may benefit from the higher level of retained austenite present in the '200' variant. A worthwhile extension of this work would be to examine the effect of tempering on other mechanical properties.

A compound layer and diffusion zone formed on all material variants (see micrographs in Figure 5a–d). The compound layers were similar in thickness and nitrogen concentration for all microstructures; however, the total nitrogen diffusion depth and concentration varied. The bainitic steel produced diffusion zones with higher nitrogen content and increased depth relative to the martensitic and pearlitic variants (Figure 5). These increased nitrogen levels and depths were reflected in the hardness profiles (Figure 5). It is evident from the very high hardness achieved (>1300 HV) in the compound layer and adjacent diffusion zone that the composition of this bainitic steel is appropriate as a nitrocarburising steel.

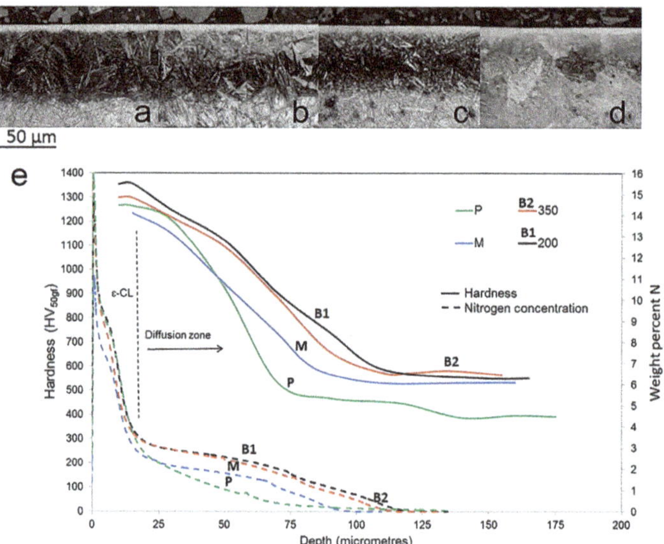

Figure 5. Nitrocarburised microstructures of (**a**) '200' bainite (B1), (**b**) '350' bainite (B2), (**c**) 'M' (**d**) 'P' and (**e**) quantitative depth profiles of nitrogen and hardness profiles in the nitrocarburised surface of all the microstructural variants.

The variations in diffusion depth and hardness may be related to the microstructures. Materials subjected to surface nanocrystallisation processes [13–15] have shown substantially increased nitrogen diffusion rates and kinetics of compound layer formation in nitriding processes. Here grain boundaries act as diffusion channels, as the diffusion rate of nitrogen at grain boundaries is twice that of the matrix [15]. The improved diffusion depth in nanostructured bainitic steel, compared to plate/lathe martensitic and coarse pearlitic steel, may be a result of the finer grain structure.

The hardness achieved in the compound layer and diffusion zone of nitrocarburised steels depends on the precipitation of fine-scale alloy nitrides. The most effective substitutional solutes are aluminium and chromium, which form CrN, Cr_2N and AlN precipitates. The higher diffusion zone nitrogen content and hardening in the nanobainitic steel compared to martensitic and pearlitic may be related to solute (Cr, Al) distribution. The partitioning of solute elements in nanostructured bainite during tempering has been characterized by using APT [10]. Below 500 °C, there is negligible solute redistribution across phase boundaries. Above 500 °C, there is enhanced redistribution of chromium around precipitated cementite, where the levels attained approach partitioning local equilibrium (7.20 at. % Cr). Chromium mobility is required for precipitate formation, so this observed segregation of chromium is of interest to the nitrocarburising process. Tong et al. [15] have found in surface nanocrystallized lath martensitic steel that Cr_2N formed near the surface during nitriding at 400 °C, which is 150 °C lower than the typical formation temperature in coarser grained martensite, and was related to the short distance required for solutes to diffuse. It is possible that the solute diffusion distances required to form alloy nitrides is reduced in the low temperature bainite, resulting in a finer dispersion of alloy nitrides. APT analysis of solute distribution in the diffusion zone of nanobainite structures is required to confirm this.

4. Conclusions

Two low transformation temperature ultrafine and nanostructured bainites, a martensitic and a pearlitic microstructural variant of a steel with the composition 0.79C-1.5Si-1.98Mn-0.98Cr-0.24Mo-1.06Al-1.58Co wt % were studied for their nitrocarburising response. After the nitrocarburising process, the core hardness of the bainitic steels was related to the amount of retained austenite in the starting microstructure. The significant core hardening experienced by the 350 °C isothermally treated steel appears to be the result of lenticular cementite precipitation in the retained austenite lath and block structure. The ultrafine or nano-scale bainitic structure is particularly suitable for nitrocarburising treatment showing good tempering resistance of the core. Additionally this structure may have assisted the diffusion of nitrogen to produce deeper nitrogen diffusion zones. Generally hardness was increased in the nitrided low temperature bainitic structure potentially as a result of the reduced diffusion distances required to form a fine dispersion of alloy nitrides.

Acknowledgments: The authors are grateful for the financial support of the Australian Research Council (ARC) through the Australian Laureate Fellowship awarded to Peter Hodgson.

Author Contributions: Daniel Fabijanic and Peter Hodgson conceived and designed the experiments. Daniel Fabijanic, Hossein Beladi and Ilana Timokhina conducted the experiments and characterization. All authors analyzed and interpreted the data. Daniel Fabijanic and Ilana Timokhina wrote the paper.

Conflicts of Interest: The authors declare no conflict of interest.

References

1. Caballero, F.G.; Bhadeshia, H.K.D.H.; Mawella, K.J.A.; Jones, D.G.; Brown, P. Design of novel high strength bainitic steels. *Mater. Sci. Technol.* **2001**, *17*, 517–522. [CrossRef]
2. Bhadeshia, H.K.D.H. *Bainite in Steels*, 2nd ed.; Institute of Materials: London, UK, 2001.
3. Garcia-Mateo, C.; Caballero, F.G. Ultra-high-strength bainitic steels. *ISIJ Int.* **2005**, *45*, 1736–1740. [CrossRef]
4. Timokhina, I.B.; Beladi, H.; Xiong, X.Y.; Adachi, Y.; Hodgson, P.D. Nano-scale microstructural characterization of a nanobainitc steel. *Acta Mater.* **2011**, *59*, 5511–5522. [CrossRef]
5. Beladi, H.; Adachi, Y.; Timokhina, I.; Hodgson, P.D. Crystallographic analysis of nanobainitic steels. *Scr. Mater.* **2009**, *60*, 455–460. [CrossRef]
6. Beladi, H.; Tari, V.; Timokhina, I.B.; Cizek, P.; Rohrer, G.S.; Rollett, A.; Hodgson, P.D. On the crystallographic characteristics of nanobainitic steel. *Acta Mater.* **2017**, *127*, 426–437. [CrossRef]
7. Wang, T.S.; Yang, J.; Shang, C.J.; Li, X.Y.; Lv, B.; Zhang, M.; Zhang, F.C. Sliding friction surface microstructure and wear resistance of 9SiCr steel with low-temperature austempering treatment. *Surf. Coat. Technol.* **2008**, *202*, 4036–4040. [CrossRef]
8. Narayanaswamy, B.; Hodgson, P.; Timokhina, I.; Beladi, H. The impact of retained austenite characteristics on the two-body abrasive wear behavior of ultra-high strength bainitic steels. *Metall. Mater. Trans.* **2016**, *47*, 4883–4895. [CrossRef]
9. Caballero, F.G.; Miller, M.K.; Clarke, A.J.; Garcia-Mateo, C. Examination of carbon partitioning into austenite during tempering of bainite. *Scr. Mater.* **2010**, *63*, 442–445. [CrossRef]
10. Caballero, F.G.; Miller, M.K.; Garcia-Mateo, C.; Capdevila, C.; Babu, S.S. Redistribution of alloying elements during tempering of a nanocrystalline steel. *Acta Mater.* **2008**, *56*, 188–189. [CrossRef]
11. Budinski, K.G. *Surface Engineering for Wear Resistance*; Prentice-Hall: Upper Saddle River, NJ, USA, 1988.
12. Tong, W.P.; Tao, N.R.; Wang, Z.B.; Lu, J.; Lu, K. Nitriding iron at lower temperatures. *Science* **2003**, *299*, 686–688. [CrossRef] [PubMed]
13. Lin, Y.; Lu, J.; Wang, L.; Xu, T.; Xue, Q. Surface nanocrystallization by surface mechanical attrition treatment and its effect on structure and properties of plasma nitrided AISI 321 stainless steel. *Acta Mater.* **2006**, *54*, 5599–5605. [CrossRef]
14. Tong, W.P.; Tao, N.R.; Wang, Z.B.; Zhang, H.W.; Lu, J.; Lu, K. The formation of ε-Fe$_{3-2}$N phase in a nanocrystalline Fe. *Scr. Mater.* **2004**, *50*, 647–650. [CrossRef]

15. Tong, W.P.; Han, Z.; Wang, L.M.; Lu, J.; Lu, K. Low-temperature nitriding of 38CrMoAl steel with a nanostructured surface layer induced by surface mechanical attrition treatment. *Surf. Coat. Technol.* **2008**, *202*, 4957–4963. [CrossRef]

16. Peet, M.J.; Babu, S.S.; Miller, M.K.; Bhadeshia, H.K.D.H. Tempering of low-temperature bainite. *Metall. Mater. Trans. A* **2017**, *48*, 1–9. [CrossRef]

17. Garcia-Mateo, C.; Caballero, F.G.; Bhadeshia, H.K.D.H. Acceleration of low temperature bainite. *ISIJ Int.* **2003**, *43*, 1821–1825. [CrossRef]

18. Cullity, B.D.; Weymouth, J.W. *Elements of X-ray Diffraction*, 2nd ed.; Addison-Wesley Publishing: Boston, MA, USA, 1978; p. 555.

19. Fonda, R.W.; Spanos, G.; Vandermeer, R.A. Observations of plate martensite in a low carbon steel. *Scr. Metall. Mater.* **1994**, *31*, 683–686. [CrossRef]

20. Morito, S.; Huang, X.; Furuhara, T.; Maki, T.; Hansen, N. The morphology and crystallography of lath martensite in alloy steels. *Acta Mater.* **2006**, *54*, 5323–5331. [CrossRef]

Article

In Situ Study of Phase Transformations during Non-Isothermal Tempering of Bainitic and Martensitic Microstructures

S. Hesamodin Talebi [1,*], Hadi Ghasemi-Nanesa [1], Mohammad Jahazi [1,*] and Haikouhi Melkonyan [2]

1   Département de Génie Mécanique, École de Technologie Supériere, Montréal, QC H3C 1K3, Canada; hadighaseminanesa@gmail.com
2   Finkl Steel Inc., 100 McCarthy, Saint-Joseph-de-Sorel, QC J3R 3M8, Canada; hmelkonyan@finkl.com
*   Correspondence: hesam.talebi@gmail.com (S.H.T.); mohammad.jahazi@etsmtl.ca (M.J.); Tel.: +1-514-396-8974 (M.J.)

Received: 30 July 2017; Accepted: 1 September 2017; Published: 4 September 2017

Abstract: Phase transformations during non-isothermal tempering of bainitic or martensitic microstructures obtained after quenching of a medium-carbon low-alloy steel was studied. The microstructures correspond to different locations of an as-quenched large-sized forged ingot used as a die material in the automotive industry. High-resolution dilatometry experiments were conducted to simulate the heat treatment process, as well as to investigate different phenomena occurring during non-isothermal tempering. The microstructures were characterized using optical and scanning electron microscopy. Dilatometry analyses demonstrated that tempering behavior varied significantly from bainitic to martensitic microstructures. Retained austenite, which exists between bainitic ferrite sheaves, decomposes to lower bainite causing a remarkable volume increase. It was found that this decomposition finishes below 386 °C. By contrast, martensite tempering was accompanied with a volume decrease due to the decomposition of medium-carbon martensite to low carbon martensite and carbides.

Keywords: high strength steel; tempering; dilatation behavior; phase transformation; microstructure; bainite; martensite

---

1. Introduction

Automotive industries' high demands for large size plastic components, such as bumpers and dashboards, have resulted in the production of ever larger-sized die steels. Consistent and uniform mechanical properties throughout the volume of the die material are required by industry [1]. The die materials are generally made of medium-carbon low-alloy steels and their manufacturing process consists of ingot casting, open die forging, quenching, tempering, and final machining [1,2]. The heat treatment usually includes austenitizing in the 840–880 °C temperature range, quenching and double-tempering between 550 °C and 600 °C.

The as-quenched microstructure is comprised of bainite, martensite, plus some retained austenite [3–5]. The transformation of austenite to bainite and martensite during quenching is dependent on the cooling rate, chemical composition of the steel and prior austenite grain size and can be predicted by continuous cooling transformation (CCT) diagrams [3,6,7]. The as-quenched microstructure will then rely on the mutual influences of the above factors. However, due to the large size of the forged block, significant temperature variations are generated between the surface and the center during quenching. Therefore, as it is difficult to improve the quenched microstructure, the material properties are optimized by modification of tempering parameters [8].

Hence, systematic study of phase transformations and microstructure evolution of the die steel during non-isothermal tempering is highly desired.

Phase transformation during non-isothermal tempering with a constant heating rate can be divided into the following stages:

Stage 1 (occurring below 80 °C): The stress fields of dislocations, lattice defects, and cell walls in the martensite, particularly interstitial lattice sites close to these defects provide lower energy sites for carbon. Stage 1 of tempering consists of segregation of carbon atoms to these lower energy sites and pre-precipitation clustering in the iron matrix [4,9–13].

Stage 2 (100–200 °C): In steels containing more than 0.2 wt % C, at temperatures between 100 °C and 200 °C, ε-carbide is the first carbide precipitating in the matrix. Some authors also reported η-carbide formation in low carbon steel at this temperature range [11,14,15].

Stage 3 (200–350 °C): Retained austenite, if present, decomposes at tempering temperatures of 200 °C to 350 °C. The behavior of retained austenite during tempering significantly varies as a function of steel composition. It generally consists of transformation into ferrite and cementite (Fe_3C) [4,9,12,16]. However, transformation into lower bainite has been reported by Yan et al. [8]. Additionally, Podder and Bhadeshia's investigations on retained austenite in bainitic steels revealed that retained austenite transformed into martensite in the cooling process after tempering [17].

Stage 4 (250–450 °C): This stage commonly referred to as the final stage of tempering consists in the transformation of the metastable ε-carbide to cementite in the temperature range of 250 °C to 450 °C. The preliminary morphology of cementite is needle-like which nucleates from martensite lath boundaries or ferrite grain boundaries. This phase transformation causes a remarkable length decrease and can be clearly detected during the dilatometry test [4,9,12,18].

It is worth mentioning that the above temperature ranges are approximate values and can differ depending on steel composition, microstructure and heating rate [4,9–12]. Time temperature transformation diagrams (TTT) are also used to determine the phase transformation of quenching and tempering process in non-equilibrium conditions but they are not capable to predict different stages during non-isothermal tempering either retained austenite decomposition. Despite many research studies dealing with martensitic phase transformation during tempering [9–17,19,20], little data is available on the evolution of bainitic microstructure during tempering of medium-carbon low alloy steels particularly, during non-isothermal heating [21]. The present study seeks to contribute to this aspect by considering phase transformation in a bainitic microstructure during non-isothermal tempering with different heating rates and provides a clear understanding of different phase transformations including retained austenite decomposition during tempering of bainite. For comparison purposes, a martensitic microstructure of the same steel is taken as reference. A combination of high-resolution dilatometry and microstructure observations were used to carry out this investigation. The obtained results are critically analyzed in terms of the evolution of carbon concentration of primary microstructure and correlated with microstructural features as a result of phase transformation.

2. Materials and Methods

The chemical composition of the steel used in the investigation was (wt %) C 0.35–V 1.49–Mn 0.99–Si 0.41–Ni 0.5–Cr 1.86–Mo 0.53–Cu 0.16. The material, provided by Finkl Steel Inc. (Sorel, QC, Canada) was from an as-forged block with initial dimensions of 6000 mm × 1200 mm × 800 mm. From the as-received steel, samples were cut into 10 mm length and 4 mm diameter cylinders. Figure 1 shows the microstructure of the as-received steel consisting in bainite and 12.1% of retained austenite.

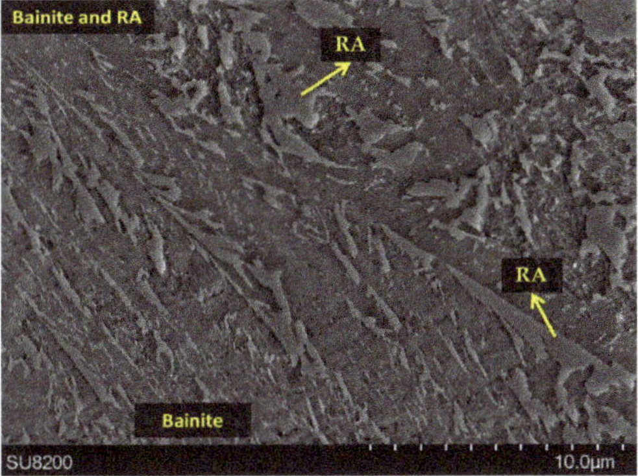

Figure 1. SEM (scanning electron microscope) image illustrating as-received microstructure composed of bainite and retained austenite (RA).

A high-resolution TA DIL 805A/D dilatometer (TA instruments, New Castle, DE, USA) with a 50-nm resolution was used to perform the heat treatment process and the dilatometry tests. Primarily, the specimens were austenitized at 870 °C for 10 min under vacuum, followed by cooling to room temperature with cooling rates of 4.8 and 180 °C/min in order to reach bainitic and a fully martensitic microstructure, respectively. To determine the optimum cooling rates, several tests using cooling rates from 2.5 °C/min to 1800 °C/min were initially carried out. Figure 2 illustrates the as-cooled martensite with M_s (Martensite start temperature) of 316 °C and bainite with B_s (Bainite start temperature) of 472 °C containing less than 5% and 23.2% of retained austenite, respectively. All the retained austenite measurements were carried out according to ASTM E975.

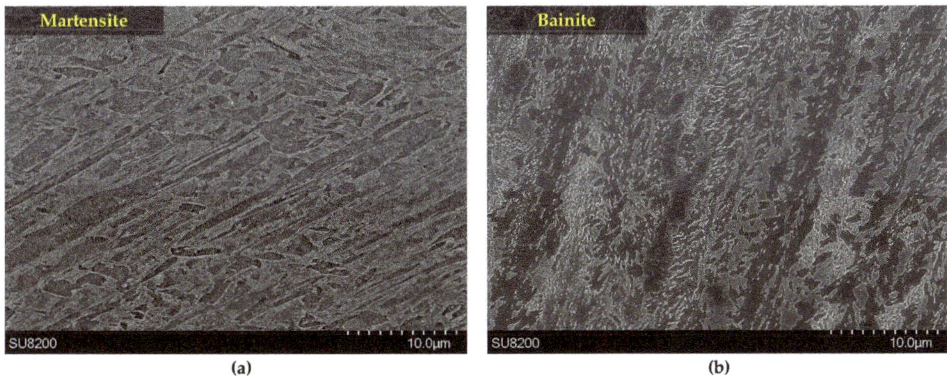

(a)   (b)

Figure 2. As-cooled microstructures obtained for different cooling rates: (**a**) Martensitic microstructure, 180 °C/min and (**b**) bainitic microstructure, 4.8 °C/min.

Subsequently, non-isothermal tempering up to 600 °C with heating rates of 5 and 30 °C/min were performed immediately after cooling to avoid possible segregation due to room temperature aging. To avoid any oxidation and decarburization, the non-isothermal tempering was conducted in the vacuum environment. After non-isothermal tempering, the samples were cooled down with a cooling rate of 600 °C/min. The relevant etchant was Vilella solution and Hitachi-SU8200 field emission gun SEM (scanning electron microscope; Hitachi, Tokyo, Japan) was used for microstructural studies. X-ray diffraction (XRD) with Co Kα radiation in standard θ–2θ mode was performed for phase identification using a Bruker Discover D8-2D diffractometer (Bruker, Madison, WI, USA). All the peaks' positions recorded by XRD were identified with the powder diffraction files (PDF) of the International Centre for Diffraction Data (ICDD) using the software Diffrac.Eva (version 4.0, Bruker AXS, Karlsruhe, Germany). Further analysis of the dilatometry results, using the first derivative of the dilatometry data, allowed for a more precise determination of the different stages of tempering. MIP image analysis software (MIP4, Nahamin Pardazan Asia, Mashhad, Iran) was used for calculating the orientation angle of the carbides within bainitic ferrite subunits.

3. Results and Discussion

Figures 3a and 4a illustrate the dilatation curves versus temperature during non-isothermal tempering related to the bainitic and martensitic specimens and Figures 3b and 4b their corresponding first derivatives. Where ΔL and L_0 are the length change and initial length of the specimen, respectively. In the following sections, dilatometry results are analyzed for each of the four stages of tempering according to the temperature range suggested for each stage.

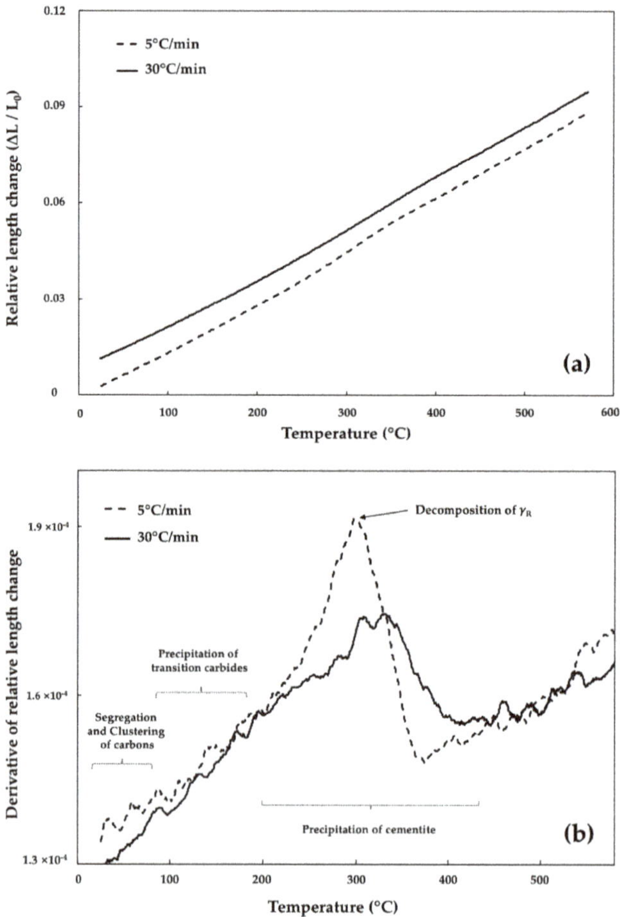

Figure 3. (a) Relative length change diagram during non-isothermal tempering of bainitic specimen at heating rates of 5 °C/min and 30 °C/min, and (b) the first derivative curve of the relative length change corresponding to these heating rates.

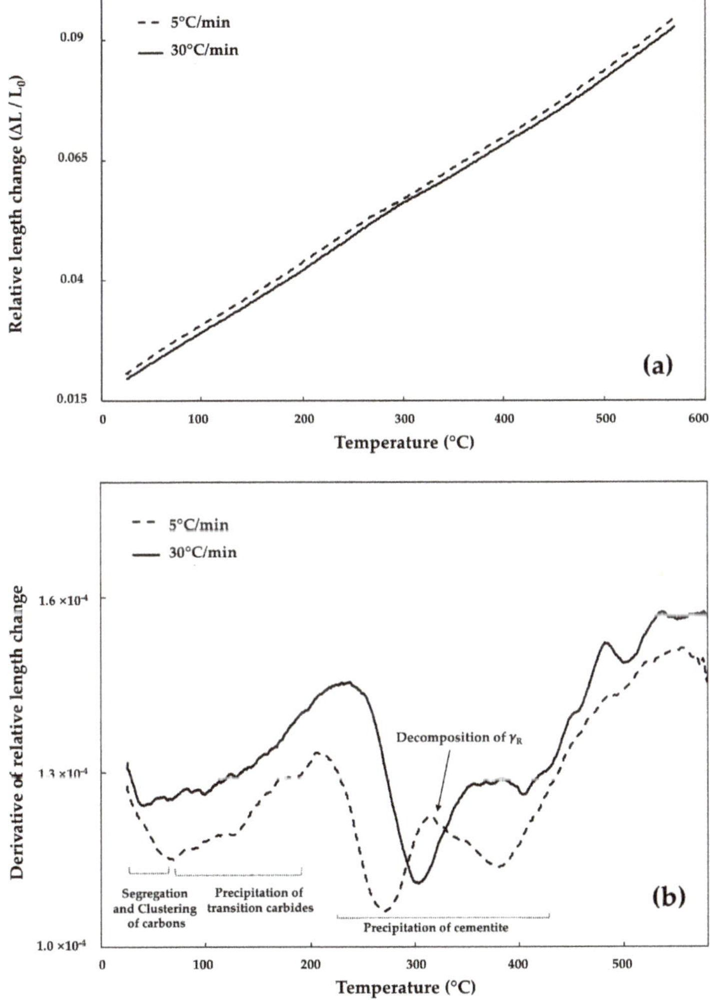

Figure 4. (**a**) Relative length change diagram during non-isothermal tempering of the martensitic specimen at the heating rates of 5 °C/min and 30 °C/min, and (**b**) the first derivative curve of relative length change corresponding to these heating rates.

3.1. Segregation and Clustering of Carbon Atoms

The first stage of tempering causes carbon atom redistribution through the BCC (Body Centered Cubic) iron with carbon segregation to the lattice defects and clustering in the matrix [22]. Examination of the derivative curve in Figure 3b did not reveal any clear indication of length changes for the two heating rates at temperatures below 80 °C. Therefore, carbon segregation and clustering within the solid solution in bainite, which should be accompanied with contractions, were not detected by dilatometry in this investigation. This could probably be related to the small carbon content of bainitic ferrite as discussed below.

In medium-carbon steels, the carbon content of martensite and bainitic ferrite (before non-isothermal tempering) can be calculated by determining the lattice parameters c and a from (110) and (200) peaks using the following equations [23]:

$$c = \frac{\lambda}{\sqrt{2}\sin\theta_{[110]}}; \; a = \frac{\lambda}{\sin\theta_{[200]}},\tag{1}$$

where λ is the wavelength of the electron beam. Therefore, the carbon content could be defined according to Equations (2) and (3) [23], where a_0 is the lattice parameter of BCC iron and $a_{0.30}$ is the lattice parameter at exactly 0.30 wt % C:

$$c = a_0 + (0.020 \pm 0.002)[C] \; ([C] < 0.56 \text{ wt \%, when } [C] < 0.30 \text{ wt \%}, c = a),\tag{2}$$

$$a = a_{0.30} - (0.014 \pm 0.002)[C] \; (0.30 \text{ wt \%} < [C] < 0.56 \text{ wt \%}).\tag{3}$$

Figure 5 illustrates XRD patterns obtained from bainitic and martensitic specimens for (110) and (200) peaks. From these peaks, lattice parameters were calculated. The lattice parameters and their related carbon contents are listed in Table 1, where [C]c and [C]a are the carbon contents based on Equations (2) and (3), respectively. According to these equations the accuracy of carbon content is ±0.02 wt %.

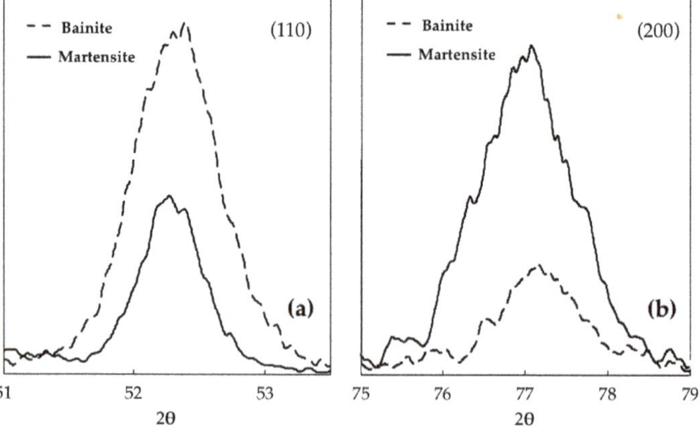

Figure 5. Two XRD (X-ray diffraction) peaks of bainite and martensite: (**a**) (110) peak and (**b**) (200) peak.

Table 1. Lattice parameters and carbon contents of the as-received microstructure, bainitic ferrite, and martensite, before and after non-isothermal tempering with a heating rate of 5 °C/min, determined using X-ray diffraction.

Microstructure	c(Å)	a(Å)	[C]c (wt %)	[C]a (wt %)	[C] (wt %)
As-received (Bainite)	2.8712	2.8668	0.24	0.30	0.26
Bainite	2.8718	2.8650	0.26	0.11	0.19
Tempered Bainite (5 °C/min)	2.8713	2.8634	0.24	0.22	0.23
Martensite	2.8724	2.8702	0.30	0.27	0.29
Tempered Martensite (5 °C/min)	2.8708	2.8689	0.21	0.17	0.19

The calculated carbon content for bainitic ferrite is 0.19 wt % before tempering and 0.23 wt % after tempering while, in martensite, this amount before and after non-isothermal tempering is 0.29 wt % and 0.19 wt %, respectively. These results are in agreement with those reported by Bhadeshia et al. [24] restating that during cooling, most of the available carbon in bainite participated in the formation of cementite particles and a small amount remained in solid solution. The low amount of carbon probably is not enough to show detectable contraction when segregation and clustering take place.

In contrast with the bainitic case, a significant contraction of martensite occurred immediately after starting the tempering (Figure 4). This behavior clearly indicates the occurrence of carbon segregation and clustering below the 80 °C temperature range in a martensitic microstructure. The higher concentration of carbon in martensite, compared to bainite, is probably the main cause for such observation and detection of the contraction by the high-resolution dilatometer during this stage of tempering.

3.2. Precipitation of the ε/η Transition Carbides

The high-resolution dilatometry results presented above (Figure 3), do not demonstrate any length decrement corresponding to ε-carbide precipitation during tempering of bainite. The amount of carbides in bainite and, therefore, carbon content of supersaturated bainitic ferrite, depends on steel composition, for instance, high silicon content delays carbide precipitation and leads to supersaturated bainitic ferrite [25]. The ε-carbide precipitation during tempering of bainite has been reported by Caballero et al. for a steel containing 1.46 wt % Si using atom probe field ion microscopy [26]. However, in the present case, the bainitic ferrite of the investigated steel is not supersaturated in carbon (0.19 wt % C) and, therefore, the precipitation of transition carbides is not expected. Speich and Leslic [11] reported that during the bainitic reaction there is a brief period for carbon redistribution and carbide precipitation called auto-tempering. Thus, the precipitation of transition carbides could presumably occur in the present steel during the formation of bainite and auto-tempering.

Contrary to tempered bainite, this stage is observed during tempering of martensite (Figure 4) because of the higher carbon amount (0.29 wt % C) in martensite. As reported in Figure 4, the start of phase transformation (length reduction) corresponding to the transition carbides overlap with the preceding phase transformation. It can also be seen that, when the higher heating rate of 30 °C/min is applied, the finish temperature is increased by 47 °C, reaching 217 °C. Jack [27] and Nakamura et al.'s [28] characterizations of the transition carbides demonstrated that tempering of martensite leads to the precipitation of ε and η transition carbides in the primarily martensitic phase.

3.3. Retained Austenite Decomposition

The decomposition of retained austenite occurs with an increase in length. Based on XRD measurements, the bainitic specimen contains 23.2% of retained austenite. Dilatometry results revealed a broad peak related to the decomposition of retained austenite that spans from 255 °C to approximately 355 °C for the heating rate of 5 °C/min. For the martensitic structure, this peak falls completely within the cementite precipitation zone and occurs in the range of 269 °C to 401 °C. It is worth noting that the XRD patterns did not show any austenite peaks in the martensitic sample, indicating the absence or a presence of less than 5% (detection limit of XRD) of retained austenite. Comparison between Figures 3 and 4 shows a clear shift of the decomposition peaks to higher temperatures, respectively 21 °C and 31 °C for bainite and martensite.

Table 2 summarizes the temperature range and relative length changes corresponding to the high and low heating rates. The degree of relative length increase is associated with the cooling rate and retained austenite percentage. These relative length changes during non-isothermal tempering are determined near the inflection points of dilatometry curves using Equation (4) [20]:

$$\text{Relative length change} = \frac{(\Delta l(T)/l)_{\text{end}} - (\Delta l(T)/l)_0}{(\Delta l(T)/l)_0}, \tag{1}$$

where $(\Delta l(T)/l)_{\text{end}}$ and $(\Delta l(T)/l)_0$ are the increase for relative length of the end and start points at temperature T, as it is represented schematically in Figure 6.

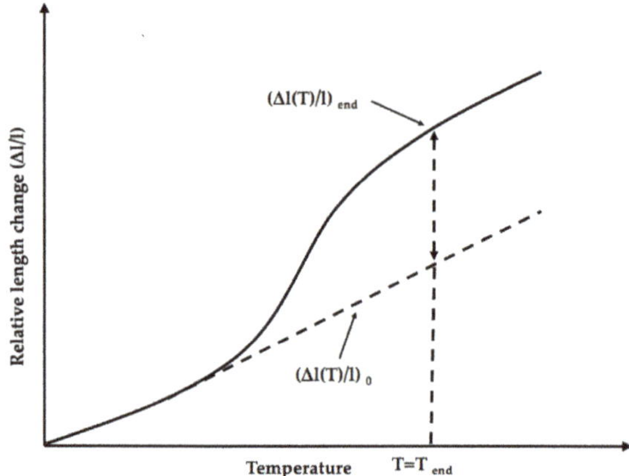

Figure 6. Relative length variations near the inflection point.

It can be seen from the data that the extent of decomposed retained austenite is associated with the heating rate. The relative volume increase corresponding to the bainitic specimen is 3.74% and 2.21% after non-isothermal tempering at the heating rate of 5 °C/min and 30 °C/min, respectively. On the other hand, the relative volume change is approximately constant in the martensitic specimen for both heating rates. These results point out that since the bainitic specimen contains a large amount of retained austenite, the portion of decomposed retained austenite has been influenced by the heating rate; therefore, after tempering at the heating rate of 30 °C/min still some non-decomposed retained austenite remains in the microstructure. In contrast, because the amount of retained austenite is very low in the martensitic structure, decomposition is almost finished even at the heating rate of 30 °C/min and no impact is observed on the relative length change value.

Table 2. Temperature interval of retained austenite decomposition and percentage of relative length change for maximum and minimum heating rates in both investigated microstructure.

Specimen	Heating Rate (°C/min)	Temperature Range (°C)	Relative Length Change (%)
Bainite	5	255–359	+3.74
	30	276–386	+2.21
Martensite	5	269–370	+0.67
	30	300–401	+0.63

3.4. Cementite Precipitation

The shape of the peak at temperature range related to the cementite precipitation varies with the microstructure. In Figure 3, there is no contraction between about 250 °C and 450 °C. Therefore, cementite precipitation was not observed for bainite. On the other hand, at the same temperature range, an intense contraction was observed in martensite associated with cementite precipitation (Figure 4). This phenomenon overlapped with retained austenite decomposition peak.

Good agreement was found with XRD results presented in Figures 7 and 8. In Figure 7, the presence of chromium carbide (Cr_7C_3) and some retained austenite are demonstrated in the bainite. Upon non-isothermal tempering at a heating rate of 5 °C/min, Fe_3C and Cr_7C_3 carbides precipitate as a consequence of retained austenite decomposition.

In the case of martensite (Figure 8), no carbides can be seen in martensite, whereas the presence of carbides in tempered martensite is evidence of a large dilatation decrease during non-isothermal tempering depicted in Figure 4b. The peak related to cementite is seen after tempering for both specimens, since cementite phase is detectable by XRD, 5% or more cementite exists in tempered steels. These Fe_3C carbides, dissolve upon tempering to form Cr_7C_3 carbides at the same location [29].

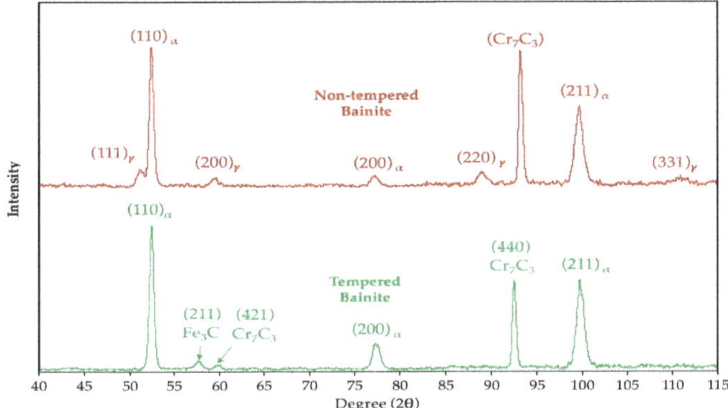

Figure 7. XRD spectra of bainite and non-isothermal tempered bainite at heating rate of 5 °C/min.

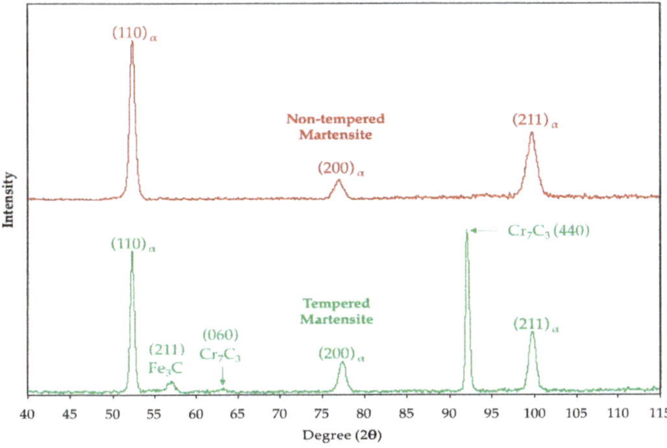

Figure 8. XRD spectra of martensite and non-isothermal tempered martensite at heating rate of 5 °C/min.

As discussed earlier, in bainite, the bainitic ferrite contains 0.19 wt % of carbon. Since there is a small amount of carbon in solid solution, the cementite precipitation during non-isothermal tempering is negligible [24]. By contrast, in supersaturated martensite with 0.29 wt % of carbon, this phase transformation takes place upon tempering between about 210 °C and 480 °C with a considerable length decrease. Similar observation for cementite precipitation in martensite was reported by Morra et al. [4]. The gradual shift of the transformation temperature to the higher values with increasing the heating rates is due to the time and temperature dependency of carbon diffusion into lattice defects.

It is also worth noting that after completing this non-isothermal tempering and cooling to the ambient temperature, final relative length, calculated by Equation (4), dropped by 0.65% in the martensitic sample after tempering, whereas by contrast, it rises by 0.44% in the bainitic specimen.

Analyses of cooling cycles after each non-isothermal tempering are represented in Figure 9 to investigate the formation of new phases during cooling. In Figure 9a, it can be seen that the first derivation of the relative length change is almost constant during cooling. The pattern is similar for the other three cooling cycles. Thus, it can be concluded that no new phase transformation occurred during the fast cooling after non-isothermal tempering.

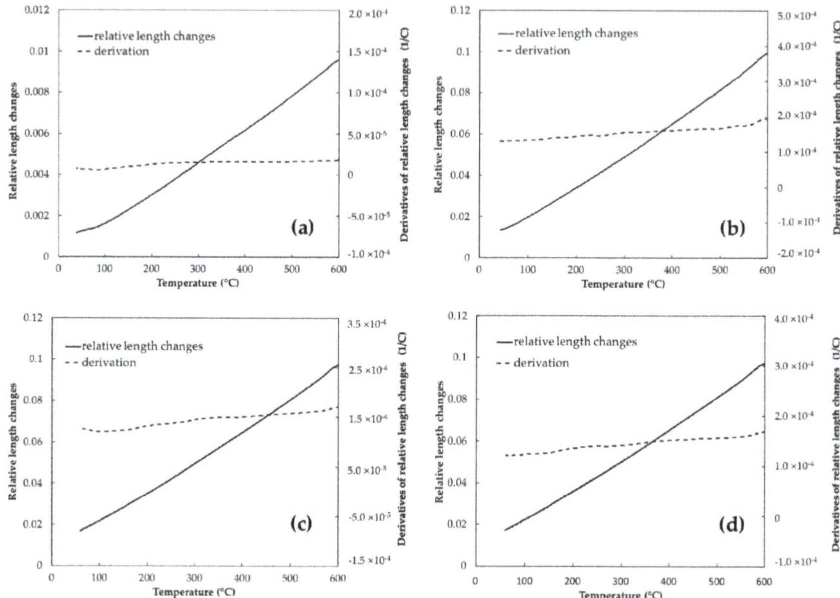

Figure 9. Relative length change and its derivation during the cooling cycle (from 600 °C to ambient temperature at a cooling rate of 600 °C/min) after tempering of: (**a**) The bainitic specimen non-isothermally tempered at a heating rate of 5 °C/min; (**b**) the bainitic specimen non-isothermally tempered at a heating rate of 30 °C/min; (**c**) the martensitic specimen non-isothermally tempered at a heating rate of 5 °C/min and (**d**) the martensitic specimen non-isothermally tempered at a heating rate of 30 °C/min.

3.5. Microstructural Evolution during Tempering

In order to study the detailed morphology after tempering, SEM observation was performed for both martensitic and bainitic specimens, before and after non-isothermal tempering. Figure 10a shows SEM images of tempered bainite after heating at the rate of 5 °C/min. Examination of the microstructure before tempering revealed the presence of several retained austenite islands with blocky morphologies at the grain boundaries of the bainitic ferrites and are characterized by a smooth surface under SEM (Figure 10b). As discussed in Section 3.3, these retained austenite islands decompose in the range of 255 °C to 355 °C and as reported in Figure 10c, their smooth surface becomes roughened with the presence of very fine carbides distributed regularly over the surface. According to Yan et al. [8], the decomposition of retained austenite may lead to the formation of martensite or lower bainite depending on the applied temperature and holding time during tempering.

In Figure 10c, the small carbides ordered in parallel bands among bainitic ferrite subunits were observed at the grain boundaries of bainitic ferrite. The axes of these carbides were inclined between 53° and 62° to the growth direction of the bainitic ferrite. It has been reported that the angle between carbides in the lower bainite ferrite inclines at about 60° to the growth direction of ferrite, which is very close to the angles measured in the present investigation [30]. It should also be noted that in agreement with reported results in the literature, the morphology of decomposed retained austenite blocks is very similar to that of lower bainite [31]. On the basis of the above analysis, it can, therefore, reasonably said that the decomposition of retained austenite results in the formation of lower bainite in the investigated steel.

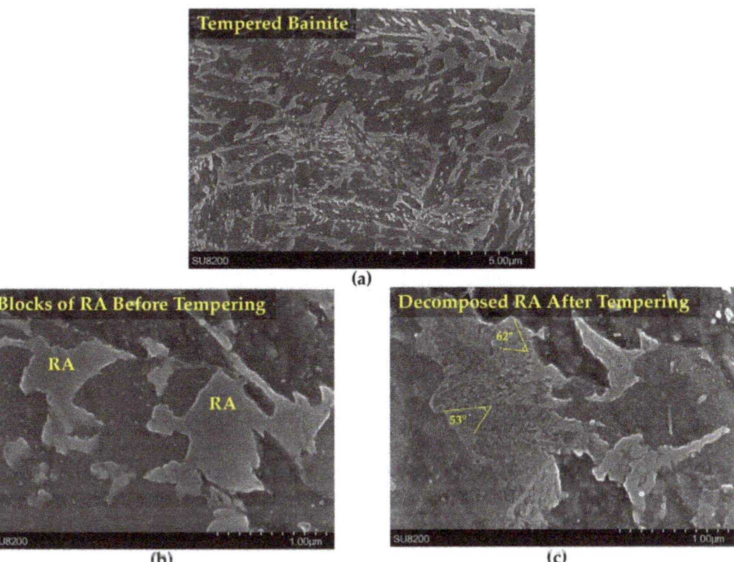

Figure 10. SEM images of microstructure evolution in the bainitic sample during non-isothermal tempering at heating rate of 5 °C/min: (**a**) The microstructure after tempering; (**b**) blocks of retained austenite (RA) prior to tempering and (**c**) blocks of decomposed retained austenite after tempering.

The tempered martensitic and bainitic microstructures at heating rates of 30 °C/min are illustrated in Figure 11. Figure 11a shows that the microstructure of tempered martensite consists in martensite plates with fine needle-like carbides, composed of cementite and chromium carbide, within the plates. Analyses of tempered bainite (Figure 11b), shows the presence of coarse rod-shape carbides. This indicates carbide *coarsening* rather than *precipitation* has taken place during non-isothermal tempering of bainite. Since most of carbides have been formed during bainitic transformation; they have, therefore, more time to grow during non-isothermal tempering.

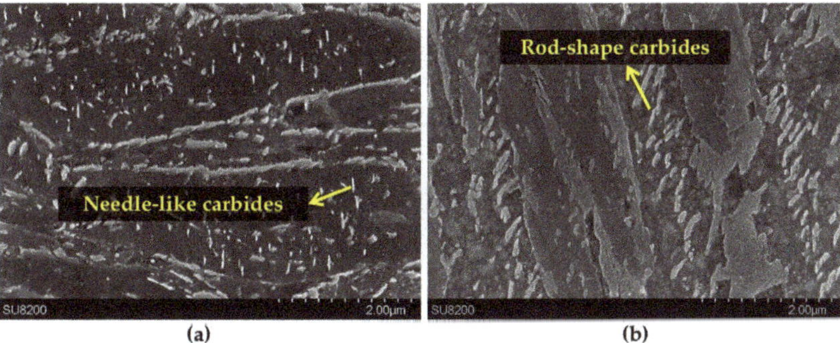

Figure 11. SEM images after non-isothermal tempering at heating rate of 30 °C/min: (**a**) Plates of tempered martensite with fine needle-like carbides within the plate and (**b**) tempered bainite with coarser rod-shape carbides.

4. Conclusions

The non-isothermal tempering behavior of a bainitic microstructure of a medium-carbon high-strength steel was investigated for two different heating rates. A martensitic structure was also used as a reference for comparison purposes. Phase transformations were studied using high-resolution dilatometry and the microstructural features were examined using optical and scanning electron microscopes. The results indicated that tempering effects vary significantly from bainitic to martensitic microstructures helping to draw a clearer picture of the phase transformations in different microstructures. The following conclusions can be drawn from the present study:

(1) In the bainitic microstructure, retained austenite decomposed to lower bainite associated with volume expansion during tempering in contrast to martensite tempering where great length decrease occurred due to decomposition of medium-carbon martensite to low carbon martensite plus carbides.

(2) Phase transformations during tempering of martensite occurred at slightly higher temperatures than bainite tempering, owing to the auto-tempering effect through bainite formation.

(3) In tempering of the bainitic specimens which contain 23.2% of retained austenite, its decomposition at the heating rate of 5 °C/min caused 3.74% length increase. This percentage decreased steeply to 2.21% when the heating rate increased to 30 °C/min, demonstrating that some amount of retained austenite remained untransformed after non-isothermal tempering at this heating rate, whilst in martensite, the length increase for both heating rates is almost constant indicating completed transformation.

Acknowledgments: The authors are very grateful to Finkl Steel Co. for providing the samples of the present research. The authors would like to appreciate the National Science and Engineering Research Council, Ottawa, ON, Canada for their financial support in the framework of a Collaborative Research and Development project (CRDJP 453683).

Author Contributions: A first draft of the manuscript was written by S. Hesamodin Talebi, which received substantial review from other authors. S. Hesamodin Talebi designed the study. S. Hesamodin Talebi and Hadi Ghasemi-Nanesa performed the experiments and analyzed the results. Mohammad Jahazi is the main supervisor and contributed to analyzing the results and writing the manuscript. Haikouhi Melkonyan is the industrial researcher at Finkl Steel, and prepared the material and evaluated the manuscript.

References

1. Firrao, D.; Matteis, P.; Russo Spena, P.; Gerosa, R. Influence of the microstructure on fatigue and fracture toughness properties of large heat-treated mold steels. *Mater. Sci. Eng. A* **2013**, *559*, 371–383. [CrossRef]

2. Firrao, D.; Gerosa, R.; Ghidini, A.; Matteis, P.; Mortarino, G.; Pinasco, M.R.; Rivolta, B.; Silva, G.; Stagno, E. Relation between fatigue crack initiation and propagation, toughness and microstructure in large steel blooms for automotive plastic molds. *Int. J. Fatigue* **2007**, *29*, 1880–1884. [CrossRef]

3. Chentouf, S.M.; Jahazi, M.; Lapierre-Boire, L.P.; Godin, S. Characteristics of austenite transformation during post forge cooling of large-size high strength steel ingots. *Metallogr. Microstruct. Anal.* **2014**, *3*, 281–297. [CrossRef]

4. Morra, P.; Böttger, A.; Mittemeijer, E. Decomposition of iron-based martensite. A kinetic analysis by means of differential scanning calorimetry and dilatometry. *J. Therm. Anal. Calorim.* **2001**, *64*, 905–914. [CrossRef]

5. Biss, V.; Cryderman, R.L. Martensite and retained austenite in hot-rolled, low-carbon bainitic steels. *Metall. Mater. Trans. B* **1971**, *2*, 2267–2276. [CrossRef]

6. Gojić, M.; Sućeska, M.; Rajić, M. Thermal analysis of low alloy Cr-Mo steel. *J. Therm. Anal. Calorim.* **2004**, *75*, 947–956. [CrossRef]

7. Rodrigues, P.C.M.; Pereloma, E.V.; Santos, D.B. Mechanical properities of an HSLA bainitic steel subjected to controlled rolling with accelerated cooling. *Mater. Sci. Eng. A* **2000**, *283*, 136–143. [CrossRef]

8. Yan, G.; Han, L.; Li, C.; Luo, X.; Gu, J. Characteristic of retained austenite decomposition during tempering and its effect on impact toughness in SA508 Gr.3 steel. *J. Nucl. Mater.* **2017**, *483*, 167–175. [CrossRef]

9. Cheng, L.; Brakman, C.M.; Korevaar, B.M.; Mittemeijer, E.J. The tempering of iron- carbon martensite; dilatometric and calorimetric analysis. *Metall. Trans. A* **1988**, *19*, 2415–2426. [CrossRef]
10. Nagakura, S.; Hirotsu, Y.; Kusunoki, M.; Suzuki, T.; Nakamura, Y. Crystallographic study of the tempering of martensitic carbon steel by electron microscopy and diffraction. *Metall. Trans. A* **1983**, *14*, 1025–1031. [CrossRef]
11. Speich, G.R.; Leslie, W.C. Tempering of steel. *Metall. Trans.* **1972**, *3*, 1043–1054. [CrossRef]
12. Primig, S.; Leitner, H. Separation of overlapping retained austenite decomposition and cementite precipitation reactions during tempering of martensitic steel by means of thermal analysis. *Thermchim. Acta* **2011**, *526*, 111–117. [CrossRef]
13. Miller, M.; Beaven, P.; Smith, G. A study of the early stages of tempering of iron-carbon martensites by atom probe field ion microscopy. *Metall. Trans. A* **1981**, *12*, 1197–1204. [CrossRef]
14. Taylor, K.; Olson, G.; Cohen, M.; Sande, J.B.V. Carbide precipitation during stage I tempering of Fe-Ni-C martensites. *Metall. Mater. Trans. A* **1989**, *20*, 2749–2765. [CrossRef]
15. Olson, G.B.; Cohen, M. Early stages of aging and tempering of ferrous martensites. *Metall. Trans. A* **1983**, *14*, 1057–1065. [CrossRef]
16. Mittemeher, E.J.; Cheng, L.; van der Schaaf, P.J.; Brakman, C.M.; Korevaar, B.M. Analysis of nonisothermal transformation kinetics; tempering of iron-carbon and iron-nitrogen martensites. *Metall. Trans. A* **1988**, *19*, 925–932. [CrossRef]
17. Saha Podder, A.; Bhadeshia, H.K.D.H. Thermal stability of austenite retained in bainitic steels. *Mater. Sci. Eng. A* **2010**, *527*, 2121–2128. [CrossRef]
18. Ghasemi Nanesa, H.; Jahazi, M. Alternative phase transformation path in cryogenically treated AISI D2 tool steel. *Mater. Sci. Eng. A* **2015**, *634*, 32–36. [CrossRef]
19. Jung, M.; Lee, S.J.; Lee, Y.K. Microstructural and dilatational changes during tempering and tempering kinetics in martensitic medium-carbon steel. *Metall. Mater. Trans. A* **2009**, *40*, 551–559. [CrossRef]
20. Leiva, J.A.V.; Morales, E.V.; Villar-Cociña, E.; Donis, C.A.; de S. Bott, I. Kinetic parameters during the tempering of low-alloy steel through the non-isothermal dilatometry. *J. Mater. Sci.* **2009**, *45*, 418. [CrossRef]
21. Peet, M.J.; Babu, S.S.; Miller, M.K.; Bhadeshia, H.K.D.H. Tempering of low-temperature bainite. *Metall. Mater. Trans. A* **2017**, *48*, 3410–3418. [CrossRef]
22. Krauss, G. Tempering of Lath Martensite in Low and Medium Carbon Steels: Assessment and Challenges. *Steel Res. Int.* **2017**. [CrossRef]
23. Kang, M.K.; Ai, Y.L.; Zhang, M.X.; Yang, Y.Q.; Zhu, M.; Chen, Y. Carbon content of bainite ferrite in 40CrMnSiMoV steel. *Mater. Chem. Phys.* **2009**, *118*, 438–441. [CrossRef]
24. Bhadeshia, H.K.D.H. *Bainite in Steels: Transformations, Microstructure and Properties*, 2nd ed.; IOM Communications: London, UK, 2001.
25. Caballero, F.G.; Miller, M.K.; Babu, S.S.; Garcia-Mateo, C. Atomic scale observations of bainite transformation in a high carbon high silicon steel. *Acta Mater.* **2007**, *55*, 381–390. [CrossRef]
26. Caballero, F.G.; Miller, M.K.; Garcia-Mateo, C.; Capdevila, C.; Babu, S.S. Redistribution of alloying elements during tempering of a nanocrystalline steel. *Acta Mater.* **2008**, *56*, 188–199. [CrossRef]
27. Jack, K. Structural transformations in the tempering of high-carbon martensitic steels. *J. Iron Steel Inst.* **1951**, *169*, 26–36.
28. Nakamura, Y.; Nagakura, S. Structure of iron-carbon martensite in the transition state from the first to the third stage of tempering studied by electron microscopy and diffraction. *Trans. Jpn. Inst. Met.* **1986**, *27*, 842–848. [CrossRef]
29. Dépinoy, S.; Toffolon-Masclet, C.; Urvoy, S.; Roubaud, J.; Marini, B.; Roch, F.; Kozeschnik, E.; Gourgues-Lorenzon, A.F. Carbide precipitation in 2.25Cr-1Mo bainitic steel: Effect of heating and isothermal tempering conditions. *Metall. Mater. Trans. A* **2017**, *48*, 2164–2178. [CrossRef]
30. Shimizu, K.; Ko, T.; Nishiyama, Z. Transmission electron microscope observation of the bainite of carbon steel. *Trans. Jpn. Inst. Met.* **1964**, *5*, 225–230. [CrossRef]
31. Porter, D.A.; Easterling, K.E.; Sherif, M. *Phase Transformations in Metals and Alloys (Revised Reprint)*, 3rd ed.; CRC Press: Boca Raton, FL, USA, 2009.

 *metals*

Article

High-Temperature Tempered Martensite Embrittlement in Quenched-and-Tempered Offshore Steels

Hung-Wei Yen [1],*, Meng-Hsuan Chiang [1], Yu-Chen Lin [1], Delphic Chen [2], Ching-Yuan Huang [2] and Hsin-Chih Lin [1]

[1]   Department of Materials Science and Engineering, National Taiwan University, Roosevelt Road, Taipei 10617, Taiwan; niko199382@gmail.com (M.-H.C.); r04527037@ntu.edu.tw (Y.-C.L.); hclinntu@ntu.edu.tw (H.-C.L.)

[2]   Iron and Steel R&D Department, China Steel Corporation, Chung Kang Road, Kaohsiung 81233, Taiwan; 185108@mail.csc.com.tw (D.C.); 148023@mail.csc.com.tw (C.-Y.H.)

*   Correspondence: homeryen@ntu.edu.tw; Tel.: +886-2-3366-1327

Received: 31 May 2017; Accepted: 3 July 2017; Published: 6 July 2017

Abstract: Embrittlement induced by high-temperature tempering was investigated in two quenched-and-tempered offshore steels. Electron backscattering diffraction and analysis of Kernel average misorientation were applied to study the coalescence of martensite, transmission Kikuchi diffraction coupled with compositional mapping was used to characterize the martensite/austenite (M/A) phases. It is suggested that the formation of lenticular martensite along prior austenite grain boundaries or packet boundaries primarily explains the embrittlement in conventional S690Q steel, which has a higher carbon content. This embrittlement can be cured by additional heat treatment to decompose martensite into ferrite and cementite. In a newly designed NiCu steel with reduced carbon content, new lath martensite formed along interlath or inter-block boundaries of prior martensite. This microstructure is less detrimental to the impact toughness of the steel.

Keywords: tempered martensite embrittlement; lenticular martensite; offshore steels; electron backscattering diffraction; Kernel average misorientation; transmission Kikuchi diffraction

---

1. Introduction

Steels with higher strength are in demand for offshore construction due to their light weight and lower material and transportation costs. One such steel is S690Q steel plate, which is recognized as a strong, tough, and highly weldable steel for offshore structures [1]. The production process of such steels usually involves quenching to transform austenite into martensite, followed by proper tempering to obtain the required properties. This quenching-and-tempering thermal cycle also occurs during multi-pass welding of these materials. Hence, the sensitivity of a material's toughness to tempering temperature is critical for offshore steels.

It has long been known that high strength martensitic steels, which are heat-treated to reach the optimum balance of strength, ductility, and toughness, are susceptible to embrittlement during tempering [2]. In the low temperature range of 230 °C to 370 °C, tempering can lead to brittleness due to strain aging, known as blue brittleness because of the blue surface scales that form [3]. Embrittlement can also occur when martensitic steel is heat treated in the temperature range of 370 °C to 575 °C. Temper embrittlement occurs during the holding or slow cooling of alloy steels previously tempered above 600 °C. This embrittlement is believed to be caused by the segregation of phosphorus, sulfur, tin, or other impurities at prior austenite grain boundaries [4,5]. Tempered martensite embrittlement occurs during tempering of as-quenched alloy steels in the range of 370 °C to 575 °C. It was originally suggested that the thermal instability of retained austenite during tempering produced embrittlement due to its decomposition to interlath films of M_3C carbides [6–8]. However, it was later found that embrittlement is concurrent with the interlath precipitation of cementite during initial tempering and the consequent mechanical instability of interlath films of retained austenite during subsequent loading [9]. Also, another study suggested that tempered martensite embrittlement is controlled by coarsening of the comparatively larger amount of interlath cementite resulting from thermal decomposition of the interlath retained austenite [10]. It is generally recognized that the occurrence of tempered martensite embrittlement cannot be attributed to a single mechanism, such as interlath carbide precipitation, decomposition of retained austenite, or impurity segregation, and that it is almost certainly due to a combination of several factors.

Temper embrittlement and tempered martensite embrittlement are usually discussed in relation to medium-carbon alloyed steels. The current work describes embrittlement due to tempering at the high temperature range of 600 °C to 780 °C in S690Q and a newly designed steel, both of which are low-carbon steels. They show strong contrasts in sensitivity to this embrittlement after high-temperature tempering. The results and discussion in this work should be useful in the alloy design of advanced offshore steels.

2. Materials and Methods

The chemical compositions of the studied steels are listed in Table 1. One is the commercially named S690Q steel, which is quenched-and-tempered (Q&T) martensitic steel. The other, named NiCu steel in this work, is a new Q&T steel having lower carbon content but higher nickel and copper contents. Both steels were produced by vacuum induction melting (VIM). The ingots were held at 1200 °C for 2 h and then cast into 160 mm-thick slabs, followed by air cooling. The slabs were thermo-mechanically processed and directly quenched in a pirate mill at China Steel Corporation. The slabs were reheated at 1200 °C for 1 h, hot rolled into 30 mm-thick plates, and then directly water quenched. The start-rolling temperature was about 1100 °C, and the finish-rolling temperature was about 800 °C. The as-quenched steels were cut into several 11 mm × 11 mm square rods. These rods were tempered at 600 °C, 660 °C, 720 °C, and 780 °C for 1 h before being quenched in a salt bath. After the heat treatments, the rods were machined into specimens for Charpy impact tests (Model IT406, Tinius Olsen, Redhill, UK). The notches were located on the plane normal to the transverse direction (TD). The testing temperature was −40 °C.

Table 1. The chemical compositions of studied offshore steels (in wt %).

	Fe	C	Si	Mn	Cr	Ni	Mo	Cu	Al	Ti + Nb	N
S690Q [1]	Bal.	0.08–0.16	0.20–0.40	0.90–1.00	<0.65	0.80–1.00	<0.50	0.10–0.40	<0.08	<0.04	<0.009
	Fe	C	Si	Mn	Cr	Ni	Mo	Cu	Al	Ti + Nb + V	N
NiCu [2]	Bal.	0.04–0.08	0.20–0.40	0.90–1.00	<0.65	2.5–3.5	<0.50	1.0–2.0	<0.08	<0.06	<0.009

[1] S ~0.0028 wt % and P ~0.01 wt %; [2] S ~0.0014 wt % and P ~0.01 wt %.

As-quenched and tempered specimens were investigated by electron backscattering diffraction (EBSD) and transmission Kikuchi diffraction (TKD) in a scanning electron microscope (SEM). EBSD experiments were conducted with an FEI NOVA SEM (FEI, Hillsboro, OR, USA) equipped with a TSL EBSD detector, and the results were processed in the Orientation Imaging Microscopy (OIM[TM]) software (version 6.2, EDAX, Mahwah, NJ, USA). TKD experiments coupled with X-ray energy-dispersive spectrum (EDS) mapping were conducted with a JSM 7800F PRIME SEM (JEOL, Tokyo, Japan) equipped with an Oxford Nordlys EBSD detector and an Oxford X-Max[50] EDS detector. The TKD results were processed in the Aztec 3.0 system and the HKL Channel 5 software. The fine structures of the M/A phase were confirmed with an FEI TECNAI F20 transmission electron microscope (TEM). Specimens for TEM, EBSD, and TKD were prepared by electrochemical polishing at 5 °C in an electrolyte mixture of 5% perchloric acid, 15% glycerol, and 80% alcohol (in vol %).

3. Results

3.1. Quenched and Tempered Microstructure

The microstructures of the as-quenched S690Q and NiCu steels are shown in Figure 1a,b, respectively. Although both steels were thermo-mechanically processed under similar parameters, the S690Q steel had equaixed prior austenite, and the NiCu steel had pancaked prior austenite. The retarded recrystallization in NiCu may have resulted from the higher contents of microalloying elements, as listed in Table 1. After water-quenching, the microstructures were lath martensite in both steels.

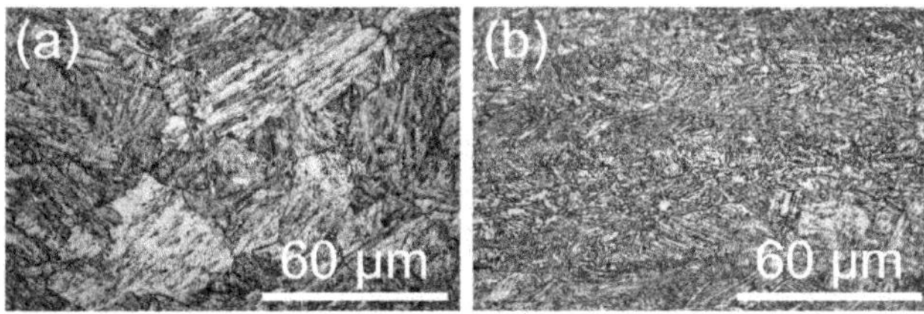

Figure 1. The optical metallography images showing the as-quenched microstructure in (**a**) S690Q steel and (**b**) NiCu steel.

The microstructures in the S690Q and NiCu steels after tempering at different temperatures are shown in Figures 2 and 3, respectively. When the tempering temperature was low, the quenched and tempered states exhibited no significant differences in microstructure, as shown in Figure 2a,b and Figure 3a,b. After tempering at 720 °C, island-like micro-phases decorated the prior austenite grain boundaries in S690Q steel, as shown in Figure 2c. Tempering at 780 °C produced a microstructure of new martensite mixed with ferrite, as shown in Figure 2d. The new martensite resulted from martensitic transformation from reverted austenite, and the ferrite formed by coalescence of old martensite. Also, the micro-phases in S690Q steel tempered at 720 °C were likely martensite, probably M/A phases. However, the fine structure was not clear under an optical microscope.

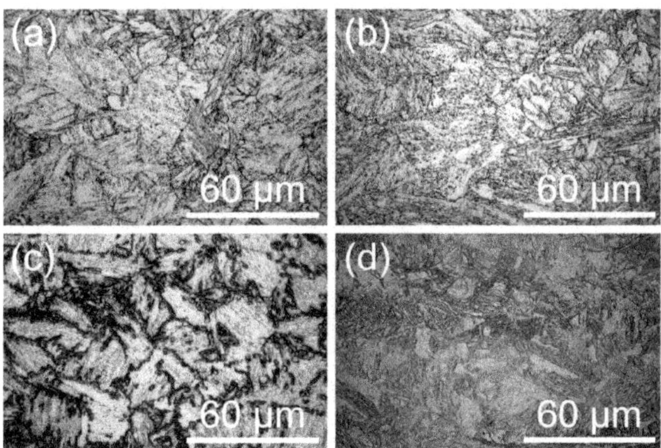

Figure 2. Optical metallography images showing the as-quenched microstructure in S690Q steel after tempering at (**a**) 600 °C; (**b**) 660 °C; (**c**) 720 °C; and (**d**) 780 °C for 1 h.

In NiCu steel tempered at 720 °C, micro-phases formed along the interlath boundaries of martensite, as shown in Figure 3c. However, the fine structure could not be clearly observed under an optical microscope. After tempering at 780 °C, no trace of pancaked prior austenite grain boundaries was visible in NiCu steel, as shown in Figure 3d. In this case, the tempering temperature was so high that the microstructure became almost fully austenite. Hence, the reverted equaixed austenite eliminated the traces of pancaked prior austenite that formed during the thermo-mechanical process.

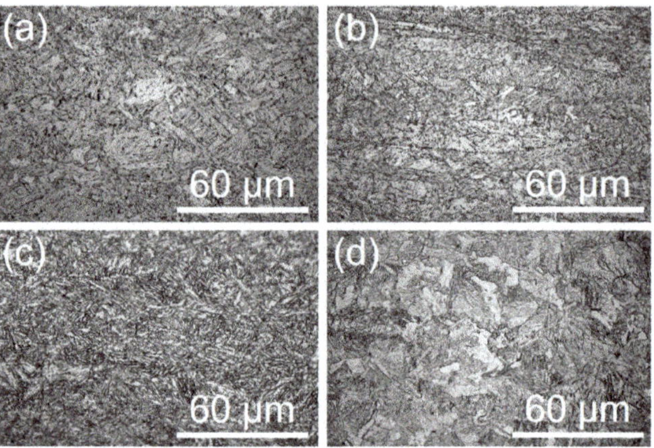

Figure 3. Optical metallography images showing the as-quenched microstructure of NiCu steel after tempering at (**a**) 600 °C; (**b**) 660 °C; (**c**) 720 °C; and (**d**) 780 °C for 1 h.

3.2. Mechanical Properties of Offshore Steels

Figure 4a shows the variation in Vickers hardness as a function of tempering temperature for both offshore steels. When tempering at low temperature, supersaturated carbon solutes partitioned into cementite, and, moreover, density of dislocation in martensite also decreased. Both factors caused the decrease in hardness of tempered martensite. As tempering temperature increased, the hardness of S690Q steel significantly decreased. However, tempering at 780 °C and quenching resulted in a large fraction of new martensite, and hardness increased to the level of the as-quenched state. In contrast, the decrease in hardness due to tempering can be mostly offset by precipitation hardening of copper in the NiCu steel. Hence, the hardness decreased slowly with increasing tempering temperature in the NiCu steel. Tempering at 600 °C for 1 h produced a yield strength of about 780 MPa and a total elongation of about 20% for S690Q steel. However, tempering at 660 °C for 1 h produced a yield strength of about 850 MPa and a total elongation of about 20% for NiCu steel. Both of them are well above the Norsok specification for the S690Q steel plate.

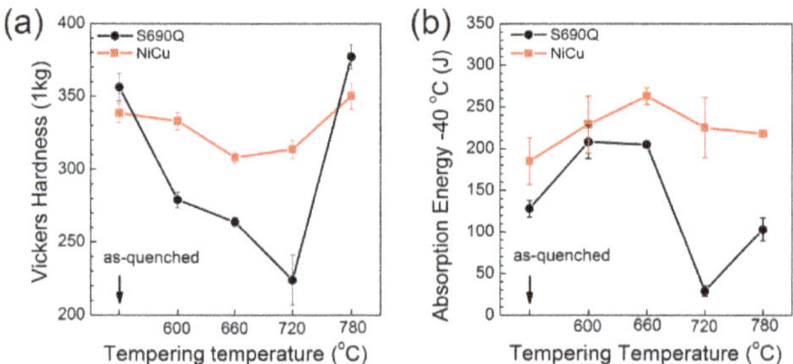

Figure 4. (**a**) Vickers hardness and (**b**) impact toughness (−40 °C) of two steels after quenching and tempering.

The absorbed energies of impact toughness at −40 °C are shown in Figure 4b for the two steels. The NiCu steel had better toughness performance. It was also found that toughness improved little, despite the elimination of pancaked prior austenite by manual quenching. Hence, the higher toughness of NiCu steel can be ascribed to its low carbon content. In most cases, tempering can improve the impact toughness of both steels. However, it is interesting that tempering at 720 °C and quenching greatly reduced the toughness of S690Q steel. The absorbed energy was only about 30 J when tested at −40 °C. In contrast, this detrimental effect was not dramatic in NiCu steel tempered at 720 °C or 780 °C. Based on the Norsok standard [1], the required impact toughness at −40 °C for 690 MPa-grade offshore steels is 100 J.

Figure 5 presents fractographs of the two steels tempered at 720 °C after the impact toughness tests at −40 °C. Consistent with the low toughness, the fractured surface of the S690Q steel revealed a brittle fracture with river cleavages. The fracture in NiCu steel was identified as ductile behavior, as indicated by the dimples.

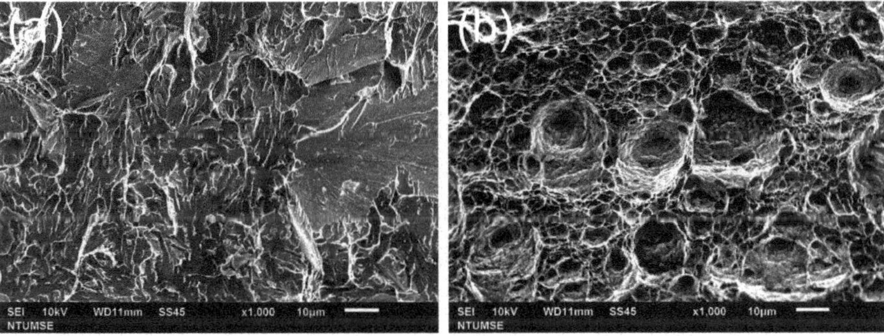

Figure 5. SEM-secondary electron images showing the fracture surfaces in the propagation regions after Charpy tests in 720 °C-tempered (**a**) S690Q steel and (**b**) NiCu steel.

3.3. Analysis of Kernel Average Misorientation (KAM)

Morphology of lath martensite in low-carbon steels has a hierarchy leveled from laths, blocks, and up to packets [11]. It has been clearly shown that variants within each sub-block always have a low-angle boundary. The strict misorientation between two variants within a sub-block is 10.53°, but the orientation distribution is scattered [12]. Besides, misorientation gradients also arise because of the plastic stains induced in austenite due to the transformation strain of martensite growth [13]. Under this circumstance, the gradual change in misorientation can be featured by the substructure of dislocations. During tempering, recovery or growth of the plate can eliminate the misoriantation, leading to coalesced martensite. It has been suggested that coalescence of martensite or bainite can have detrimental effects on the toughness of steel [14,15].

Based on EBSD results, Kernel average misorientation (KAM) is a method calculating the averaged misorientations between the center point and all surrounding points in the Kernel [16]. Hence, KAM analysis provides the local misorientation value of the center point. Coalesced martensite can be identified by applying KAM analysis because it has lower averaged values of local misorientation. In the current work, a scanning step size of 400 nm was used, and the nearest neighboring was set as the 1st order. Figure 6a–d show the KAM analyses for the tempered S690Q steel. Misorientation between martensite laths occurred even within the same martensite block. Coalescence reduces the misorientation between laths. In this analysis, the blue areas are regions of low misorientation, indicating regions corresponding to coalesced martensite. It was found that raising the tempering temperature enhanced the level of coalescence. If coalescence of martensite is the main cause of embrittlement, toughness should have decreased with increasing tempering temperature. However, absorption energy revealed an inverse trend, and the embrittlement occurred only when S690Q steel was tempered at 720 °C and quenched. Hence, it is proposed that there are other factors causing the 720 °C tempered martensite embrittlement. After tempering at 780 °C, the coalesced martensite, or ferrite, was mixed with large amounts of new martensite.

Figure 6e–h presents the KAM analyses for the quenched and tempered NiCu steel. It was found that the level of coalescence of martensite was not high in this steel. When the steel was tempered at 720 °C, very weak effects of coalescence could be observed between martensite laths, as shown in Figure 6g. For detailed observations between the martensite laths, orientation mapping was conducted at higher resolution.

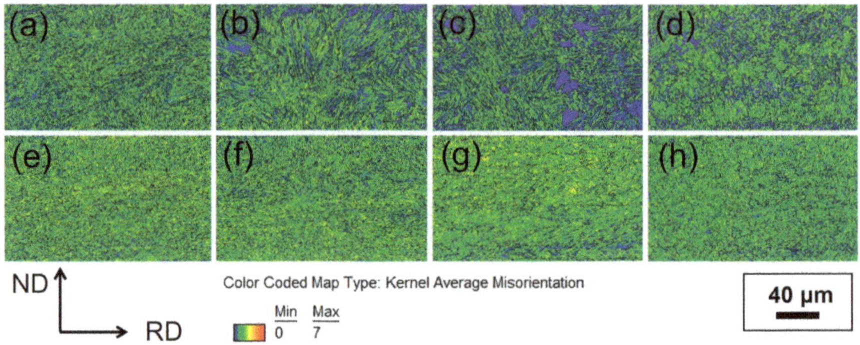

Figure 6. Kernel average misorientation (KAM) analyses of S690Q steel after tempering at (**a**) 600 °C; (**b**) 660 °C; (**c**) 720 °C; and (**d**) 780 °C for 1 h, and NiCu steel after tempering at (**e**) 600 °C; (**f**) 660 °C; (**g**) 720 °C; and (**h**) 780 °C for 1 h. RD is the rolling direction, and ND is the normal direction of the steel plate.

3.4. Investigations by Transmission Kikuchi Diffraction (TKD)

TKD can provide orientation mapping at nanoscale, and it has been applied in many studies of steels and alloys [17,18]. A scanning step size of 10 nm was used in this work. Figures 7 and 8 show the TKD analyses of M/A phases in the S690Q and NiCu steels after 720 °C tempering. In the S690Q steel, the M/A phases appeared as islands with lower band contrast (darker), as shown in Figure 7a, and they were lenticular martensite, as shown in Figure 7b. Figure 7c shows experimental and simulated {1 0 0} pole figures of martensite corresponding to the colors in Figure 7b. The simulated {1 0 0} pole figure of martensite was obtained based on Miyamoto et al.'s method [19]. It should be noted that only the ideal Kurdjumov-Sachs (KS) orientation relationship (OR) [20] was used in the analyses. There are twenty-four variants of KS OR as shown in Table 2. Based on the crystallography analysis, V13/V16, V14/V17, and V15/V18 sub-blocks belong to the same packet because they have the same {1 1 1}$_\gamma$ plane. The V20/V23 sub-block belongs to another packet. Hence, the M/A phases formed along the boundary of the packet as indicated by white arrows in Figure 7b. Besides, the martensite variants in M/A phase hold a twinning relationship (V15 and V16; V15 and V13), and they look like interlocked saws. Martensite in M/A phase is suggested to be lenticular martensite with twinned martensitic variants. Only some tiny M/A phases formed along the boundaries of sub-blocks as pointed out by black arrows in Figure 7b. The distribution of these phases is consistent with the observation in Figure 2c. Moreover, the substitutional solutes, such as Ni, Mn, and Cu, were partitioned mainly into cementite particles and partially into reverted austenite, as shown in Figure 7d–g. Also, the partitioning of carbon explains the stability of austenite at 720 °C. Austenite with higher carbon content transformed into lenticular martensite during quenching.

Figure 8 shows that, after tempering at 720 °C, M/A phases extensively formed along martensite laths in the NiCu steel. Figure 8c shows experimental and simulated {1 0 0} pole figures of martensite corresponding to the colors in Figure 8b. The variant X, which is green in Figure 8b, does belong to the same austenite. Moreover, there are three packets in the primary austenite. V14/V17, and V15/V18 sub-blocks belong to one packet, V2/V5 and V3/V6 sub-blocks belong to one packet, and V8/V11 sub-block belongs to the other. In this case, the M/A phases formed along a prior austenite boundary (pointed out by gray arrow), packet boundaries (pointed out by white arrow), and sub-block or interlath boundaries (partially pointed out by dark arrow). Besides, the martensite variants in M/A phase hold a twinning relationship (V17 and V18), but they have no saw-like feature. Also, the new martensite in M/A phase was similar to lath martensite, but its morphology was not typical due to constraint by the size of the austenite grains. The extensive formation of interlath or interblock martensite helped to suppress the coalescence of martensite. In addition, during tempering, the solutes of Ni, Mn, Cu, and C were partitioned into reverted austenite along lath boundaries, and solutes of Cr were partitioned into prior martensite in the NiCu steel. As shown in Figure 8e, some precipitates of copper were observed. TKD mapping coupled with chemical mapping showed both microstructural and chemical information of the M/A phases.

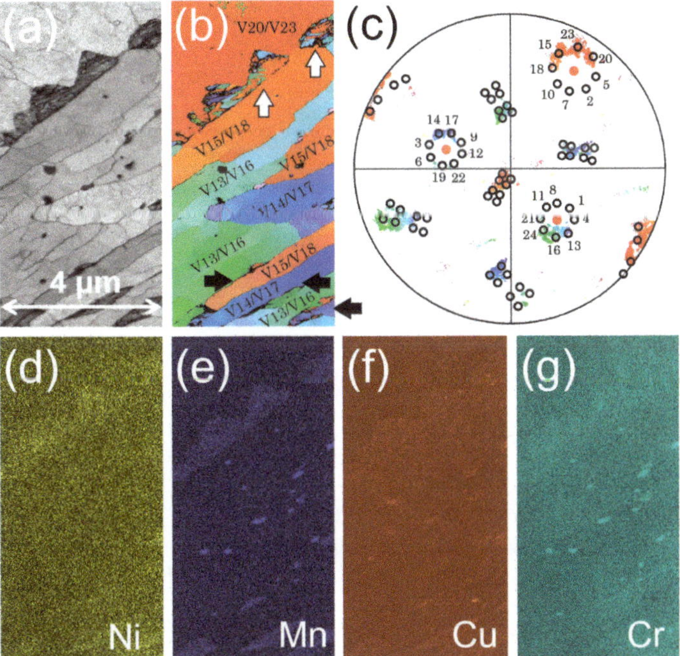

Figure 7. Transmission Kikuchi diffraction (TKD) mapping showing (**a**) band contrast; (**b**) inverse pole figure—Z; (**c**) experimental and simulated pole figures; (**d**) Ni distribution; (**e**) Mn distribution; (**f**) Cu distribution; and (**g**) Cr distribution in S690Q steel after tempering at 720 °C.

Table 2. The twenty-four crystallographic variants for Kurdjumov-Sachs (KS) orientation relationship evolved from a single austenite grain.

Variant Number	Plane Parallel	Direction Parallel
V1	$(111)_\gamma \parallel (011)_{\alpha'}$	$[\bar{1}01]_\gamma \parallel [\bar{1}\bar{1}1]_{\alpha'}$
V2	$(111)_\gamma \parallel (011)_{\alpha'}$	$[\bar{1}01]_\gamma \parallel [\bar{1}1\bar{1}]_{\alpha'}$
V3	$(111)_\gamma \parallel (011)_{\alpha'}$	$[01\bar{1}]_\gamma \parallel [\bar{1}11]_{\alpha'}$
V4	$(111)_\gamma \parallel (011)_{\alpha'}$	$[01\bar{1}]_\gamma \parallel [\bar{1}1\bar{1}]_{\alpha'}$
V5	$(111)_\gamma \parallel (011)_{\alpha'}$	$[1\bar{1}0]_\gamma \parallel [\bar{1}1\bar{1}]_{\alpha'}$

Table 2. *Cont.*

Variant Number	Plane Parallel	Direction Parallel
V6	$(111)_\gamma \parallel (011)_{\alpha'}$	$[\bar{1}10]_\gamma \parallel [\bar{1}1\bar{1}]_{\alpha'}$
V7	$(1\bar{1}1)_\gamma \parallel (011)_{\alpha'}$	$[10\bar{1}]_\gamma \parallel [\bar{1}\bar{1}1]_{\alpha'}$
V8	$(1\bar{1}1)_\gamma \parallel (011)_{\alpha'}$	$[10\bar{1}]_\gamma \parallel [\bar{1}1\bar{1}]_{\alpha'}$
V9	$(1\bar{1}1)_\gamma \parallel (011)_{\alpha'}$	$[\bar{1}10]_\gamma \parallel [\bar{1}\bar{1}1]_{\alpha'}$
V10	$(1\bar{1}1)_\gamma \parallel (011)_{\alpha'}$	$[\bar{1}10]_\gamma \parallel [\bar{1}1\bar{1}]_{\alpha'}$
V11	$(1\bar{1}1)_\gamma \parallel (011)_{\alpha'}$	$[011]_\gamma \parallel [\bar{1}\bar{1}1]_{\alpha'}$
V12	$(1\bar{1}1)_\gamma \parallel (011)_{\alpha'}$	$[011]_\gamma \parallel [\bar{1}1\bar{1}]_{\alpha'}$
V13	$(\bar{1}11)_\gamma \parallel (011)_{\alpha'}$	$[0\bar{1}1]_\gamma \parallel [\bar{1}\bar{1}1]_{\alpha'}$
V14	$(\bar{1}11)_\gamma \parallel (011)_{\alpha'}$	$[0\bar{1}1]_\gamma \parallel [\bar{1}1\bar{1}]_{\alpha'}$
V15	$(\bar{1}11)_\gamma \parallel (011)_{\alpha'}$	$[10\bar{1}]_\gamma \parallel [\bar{1}\bar{1}1]_{\alpha'}$
V16	$(\bar{1}11)_\gamma \parallel (011)_{\alpha'}$	$[10\bar{1}]_\gamma \parallel [\bar{1}1\bar{1}]_{\alpha'}$
V17	$(\bar{1}11)_\gamma \parallel (011)_{\alpha'}$	$[110]_\gamma \parallel [\bar{1}\bar{1}1]_{\alpha'}$
V18	$(\bar{1}11)_\gamma \parallel (011)_{\alpha'}$	$[110]_\gamma \parallel [\bar{1}1\bar{1}]_{\alpha'}$
V19	$(11\bar{1})_\gamma \parallel (011)_{\alpha'}$	$[\bar{1}10]_\gamma \parallel [\bar{1}\bar{1}1]_{\alpha'}$
V20	$(11\bar{1})_\gamma \parallel (011)_{\alpha'}$	$[\bar{1}10]_\gamma \parallel [\bar{1}1\bar{1}]_{\alpha'}$
V21	$(11\bar{1})_\gamma \parallel (011)_{\alpha'}$	$[01\bar{1}]_\gamma \parallel [\bar{1}\bar{1}1]_{\alpha'}$
V22	$(11\bar{1})_\gamma \parallel (011)_{\alpha'}$	$[01\bar{1}]_\gamma \parallel [\bar{1}1\bar{1}]_{\alpha'}$
V23	$(11\bar{1})_\gamma \parallel (011)_{\alpha'}$	$[101]_\gamma \parallel [\bar{1}\bar{1}1]_{\alpha'}$
V24	$(11\bar{1})_\gamma \parallel (011)_{\alpha'}$	$[101]_\gamma \parallel [\bar{1}1\bar{1}]_{\alpha'}$

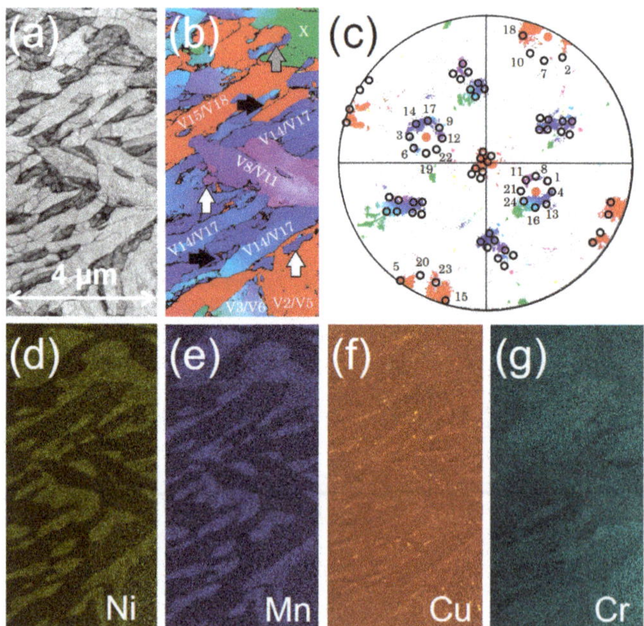

Figure 8. TKD mapping showing (**a**) band contrast; (**b**) inverse pole figure—Z; (**c**) experimental and simulated pole figures; (**d**) Ni distribution; (**e**) Mn distribution; (**f**) Cu distribution; and (**g**) Cr distribution in NiCu steel after tempering at 720 °C.

The morphology of the martensite was further confirmed by TEM. Figure 9 shows lenticular martensite in the S690Q steel after tempering at 720 °C. The lenticular martensite can be recognized by its midrib and twinned orientation relationship [21]. Figure 10 shows lath martensite in the NiCu steel after tempering at 720 °C. Analysis of the axis-angle pairs in Figure 10d revealed that the laths or blocks had a twinned relationship. This relationship occurs when an austenite grain is fine; a single martensite packet with twinned blocks/laths replaces the whole austenite [11]. This structure is sometimes described as twinned lath martensite [17]. Moreover, in all our TEM investigations, it was very difficult to find lenticular martensite in the NiCu steel. It is known here that orientation mapping from TKD is able to provide information to distinguish lenticular martensite from lath martensite in M/A phase. With further crystallography analysis, locations for reverse austenite can also be identified. Coupled with EDS mapping, structural and chemical characterizations can be done in one scanned region at one time.

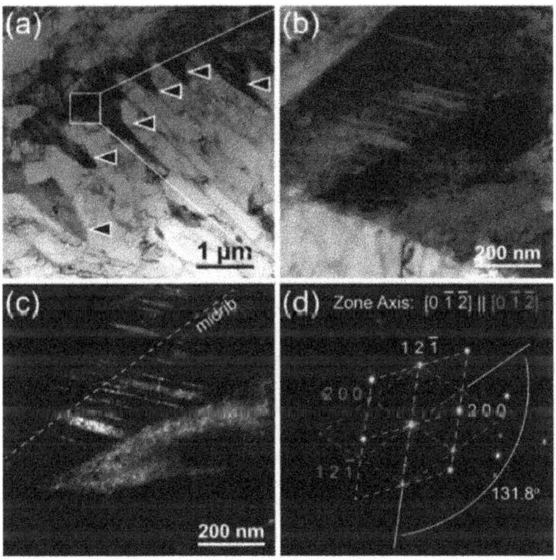

Figure 9. TEM micrographs showing the fine structure of M/A phases in S690Q steel after tempering at 720 °C. (**a**) Bright-field image at low magnification; (**b**) bright-field image; (**c**) dark-field image; and (**d**) electron diffraction pattern.

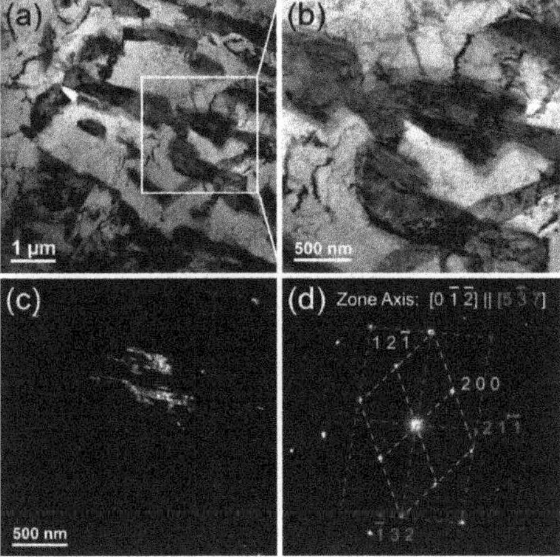

Figure 10. TEM micrographs showing the fine structure of M/A phases in NiCu steel after tempering at 720 °C. (**a**) Bright-field image at low magnification; (**b**) bright-field image; (**c**) dark-field image; and (**d**) electron diffraction pattern.

4. Discussion

This work found that embrittlement induced by high-temperature tempering was obvious in S690Q and absent in the newly-designed NiCu steel. This difference is explained by their microstructural differences, which were determined by their alloy compositions. After tempering at 720 °C, the microstructural differences can be summarized as follows:

1. The level of martensite coalescence is higher in S690Q steel;
2. Lenticular martensite forms as M/A phases in S690Q steel, whereas lath martensite forms as M/A phases in NiCu steel;
3. M/A phases form primarily along prior austenite grain boundaries or packet boundaries in S690Q steel, whereas M/A phases primarily form along lath boundaries in NiCu steel;
4. Copper particles form in NiCu steel.

The effect of martensite coalescence on this embrittlement is not significant, as discussed in Section 3.3. Although copper particles were discovered in the NiCu steel after 720 °C tempering, as shown in Figure 8f, their size is about 50–200 nm. Only a few nanometer-sized copper particles were observed, as shown in Figure 11a. When the NiCu steel tempered at 660 °C, extensive precipitation of copper particles was observed, as shown in Figure 11b. The precipitation of copper at 660 °C involves tiny intralath particles (pointed out by black arrows) with sizes of 5–20 nm and coarser interlath particles with sizes of 20–50 nm (pointed out by white arrows). After 720 °C tempering, the particle size is much coarser, and the density is extremely low. Hence, when comparing the toughness of S690Q and NiCu steels after 720 °C tempering, the difference caused by copper precipitation is suggested to be truly minor. Exactly speaking, microstructure that is free of copper precipitation in S690Q steel is not the cause of this embrittlement. Based on the above discussions, it is suggested that the high-temperature tempered martensite embrittlement in S690Q steel results primarily from the formation of lenticular martensite. To support this point, secondary tempering at 550 °C for 1 h was conducted on S690Q steel previously tempered at 720 °C. In this case, the impact toughness at −40 °C exceeded 200 J. The KAM analysis in Figure 12a shows that the level of coalescence after secondary tempering was even higher than that after only 720 °C tempering. Here, coalesced martensite was not the main factor of embrittlement. Figure 12b clearly shows that the lenticular martensite decomposed into cementite/ferrite aggregates. Hence, the formation of lenticular martensite leads to the loss in toughness in S690Q steel.

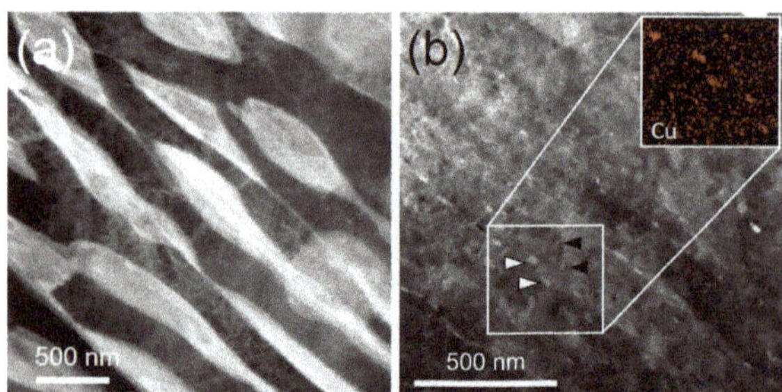

Figure 11. High-angle annular dark-field (HAADF) image showing microstructure in (**a**) S690Q steel and (**b**) NiCu steel after a 720 °C tempering for 1 h.

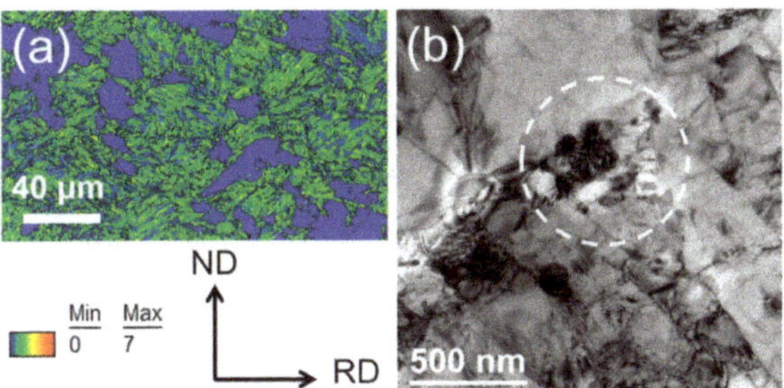

Figure 12. (**a**) KAM analysis and (**b**) TEM image showing microstructure of 690Q steel after a 720 °C tempering for 1 h, followed by secondary tempering at 550 °C for 1 h. RD is the rolling direction, and ND is the normal direction of steel plate.

Figure 13 shows the equilibrium phase fractions as a function of temperature for the two offshore steels. The results were calculated by Thermo-Calc with the TCFE8 database. The design principle of NiCu steel is to reduce the carbon content and to increase the copper content in order to obtain precipitation hardening from copper after tempering. Its higher resistance to high-temperature tempered martensite embrittlement can be understood by its phase constituents during tempering. As shown in Figure 13b, the equilibrium fraction of austenite is about 0.43 at 720 °C in the NiCu steel, and the carbon content of austenite is only about 0.09 wt %. In S690Q steel (Figure 13a), however, the equilibrium fraction of austenite is about 0.23, and the carbon content of austenite is about 0.41 wt %. Austenite with carbon content of 0.41 wt % will transform into lenticular martensite [4,11]. Moreover, the high volume fraction of new martensite, which directly forms along prior interlath boundaries, suppresses the coalescence of martensite. This microstructure also increases the toughness. Hence, in addition to the reported benefits of copper-containing high-strength low alloy steels [22,23], this alloy design shows better resistance to the high-temperature tempered martensite embrittlement that occurs in S690Q steel during welding.

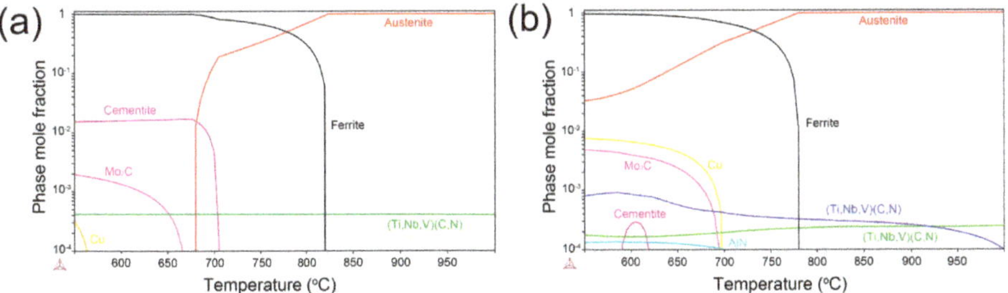

Figure 13. Equilibrium phase fractions as a function of temperature for (**a**) S690Q steel and (**b**) NiCu steel.

5. Conclusions

In this work, EBSD-KAM analysis and TKD mapping were used to investigate high-temperature tempered martensite embrittlement in steels for offshore structures. It was found that the formation of lenticular martensite during a thermal cycle of high-temperature tempering and quenching is more detrimental to the toughness of S690Q offshore steel than is the coalescence of martensite. In the newly designed nickel-copper-containing low-carbon steel, low-carbon lath martensite forms between prior martensite laths after 720 °C tempering, and this steel shows better resistance to high-temperature tempered martensite embrittlement.

Acknowledgments: This work was supported by the following projects: (1) MOST 105-2622-8-006-001 and (2) MOST 104-2218-E-002-022-MY3. The authors thank Jer-Ren Yang at the Department of Materials Science and Engineering, National Taiwan University, for his generous support in the experimental facilities of TEM sample preparation. Moreover, we specially thank Yuan-Tzu Lee for her technical support in FEG-SEM (JSM 7800F PRIME) at the Instrumentation Center, National Taiwan University, under the Ministry of Science and Technology, R.O.C.

Author Contributions: Hung-Wei Yen and Hsin-Chih Lin conceived the idea of this work and designed the materials. Delphic Chen and Ching-Yuan Huang prepared the steels and conducted the thermo-mechanical controlled processes at China Steel Corporation. Meng-Hsuan Chiang and Yu-Chen Lin conducted the heat treatments, measurements of impact toughness, and SEM characterizations. All of the authors discussed the results and reviewed the manuscript.

Conflicts of Interest: The chemical compositions of the steels were listed as ranges to prevent disclosure of the intellectual property of the China Steel Corporation. This practice does not change the main finding of this manuscript. The authors declare no conflict of interest.

References

1. Standard, N. *Material Data Sheets for Structural Steel*; Norwegian Technology Center: Oslo, Norway, 2000; Volume M-120.
2. MacKenzie, S. Overview of the mechanisms of failure in heat treated steel components. In *Failure Analysis of Heat Treated Steel Components*; Totten, G.E., Canale, L.D.C.F., Mesquita, R.A., Eds.; ASM International: Materials Park, OH, USA, 2008; p. 640.
3. Sinha, A.K. *Physical Metallurgy Handbook*; McGraw-Hill: New York, NY, USA, 2003.
4. Honeycombe, R.K.W.; Bhadeshia, H.K.D.H. *Steels: Microstructure and Properties*, 3rd ed.; Elsevier Ltd., International: Butterworth-Heinemann, Oxford, UK, 2006.
5. Briant, C.L.; Banerji, S.K. Tempered martensite embrittlement in phosphorus doped steels. *Metall. Trans. A* **1979**, *10*, 1729–1737. [CrossRef]
6. Thomas, G. Retained austenite and tempered martensite embrittlement. *Metall. Trans. A* **1978**, *9*, 439–450. [CrossRef]
7. Sarikaya, M.; Jhingan, A.K.; Thomas, G. Retained austenite and tempered martensite embrittlement in medium carbon steels. *Metall. Trans. A* **1983**, *14*, 1121–1133. [CrossRef]
8. Materkowski, J.P.; Krauss, G. Tempered martensite embrittlement in sae 4340 steel. *Metall. Trans. A* **1979**, *10*, 1643–1651. [CrossRef]
9. Horn, R.M.; Ritchie, R.O. Mechanisms of tempered martensite embrittlement in low alloy steels. *Metall. Trans. A* **1978**, *9*, 1039–1053. [CrossRef]
10. Bhadeshia, H.; Edmonds, D. Tempered martensite embrittlement: Role of retained austenite and cemenite. *Met. Sci.* **1979**, *13*, 325–334.
11. Krauss, G. *Steels: Processing, Structure, and Performance*; ASM International: Materials Park, OH, USA, 2005.
12. Kitahara, H.; Ueji, R.; Tsuji, N.; Minamino, Y. Crystallographic features of lath martensite in low-carbon steel. *Acta Mater.* **2006**, *54*, 1279–1288. [CrossRef]
13. Pak, J.H.; Bhadeshia, H.K.D.H.; Karlsson, L. Mechanism of misorientation development within coalesced martensite. *Mater. Sci. Technol.* **2012**, *28*, 918–923. [CrossRef]
14. Pous-Romero, H.; Bhadeshia, H. Coalesced martensite in pressure vessel steels. *J. Press. Vessel Technol.* **2014**, *136*, 031402–031406. [CrossRef]
15. Bhadeshia, H.K.D.H.; Keehan, E.; Karlsson, L.; Andrén, H.O. Coalesced bainite. *Trans. Indian Inst. Met.* **2006**, *59*, 6.
16. Wright, S.I.; Nowell, M.M.; Field, D.P. A review of strain analysis using electron backscatter diffraction. *Microsc. Microanal.* **2011**, *17*, 316–329. [CrossRef] [PubMed]
17. Yen, H.-W.; Ooi, S.W.; Eizadjou, M.; Breen, A.; Huang, C.-Y.; Bhadeshia, H.K.D.H.B.; Ringer, S.P. Role of stress-assisted martensite in the design of strong ultrafine-grained duplex steels. *Acta Mater.* **2015**, *82*, 100–114. [CrossRef]
18. Trimby, P.W.; Cao, Y.; Chen, Z.; Han, S.; Hemker, K.J.; Lian, J.; Liao, X.; Rottmann, P.; Samudrala, S.; Sun, J.; et al. Characterizing deformed ultrafine-grained and nanocrystalline materials using transmission kikuchi diffraction in a scanning electron microscope. *Acta Mater.* **2014**, *62*, 69–80. [CrossRef]

19. Miyamoto, G.; Takayama, N.; Furuhara, T. Accurate measurement of the orientation relationship of lath martensite and bainite by electron backscatter diffraction analysis. *Scr. Mater.* **2009**, *60*, 1113–1116. [CrossRef]
20. Kurdjumov, G.V.; Sachs, G. ber den mechanismus der stahlhärtung. *Z. Phys.* **1930**, *64*, 325–343. [CrossRef]
21. Lee, H.-Y.; Yen, H.-W.; Chang, H.-T.; Yang, J.-R. Substructures of martensite in Fe-1C-17Cr stainless steel. *Scr. Mater.* **2010**, *62*, 670–673. [CrossRef]
22. Mujahid, M.; Lis, A.K.; Garcia, C.I.; DeArdo, A.J. Hsla-100 steels: Influence of aging heat treatment on microstructure and properties. *J. Mater. Eng. Perform.* **1998**, *7*, 247–257. [CrossRef]
23. Ray, P.K.; Ganguly, R.I.; Kumar Panda, A. Influence of heat treatment parameters on structure and mechanical properties of an hsla-100 steel. *Steel Res.* **2002**, *73*, 347–355. [CrossRef]

Article

Development of Direct Quenched Hot Rolled Martensitic Strip Steels

Lieven Bracke *, Dorien De Knijf, Christoph Gerritsen, Reza Hojjati Talemi and Eva Diaz Gonzalez

ArcelorMittal Global R&D Ghent, J.F. Kennedylaan, 9060 Zelzate, Belgium;
dorien.deknijf@arcelormittal.com (D.D.K.); christoph.gerritsen@arcelormittal.com (C.G.);
reza.hojjatitalemi@arcelormittal.com (R.H.T.); eva.diazgonzalez@arcelormittal.com (E.D.G.)
* Correspondence: lieven.bracke@arcelormittal.com; Tel.: +32-477-026-046

Received: 7 June 2017; Accepted: 22 August 2017; Published: 24 August 2017

Abstract: Metallurgical concepts for new ultra-high strength martensitic steels have been developed through direct quenching after hot rolling. In addition to the chemical composition, the hot rolling, quenching, and annealing parameters need to be optimized to fulfill the requirements for the demanding applications for which these steels are used. It is also shown that the welding behavior is influenced by the choice of alloying concept. Typical applications also require a high fatigue resistance, especially of formed components. For that reason, a dedicated set-up was developed that allows differentiation between materials, which is illustrated through the effect of inclusions on the fatigue performance of a bent test piece.

Keywords: martensitic steel; direct quenched; industrialization; hot rolling; tempering; welding; fatigue

---

1. Introduction

For many years, the production of martensitic flat steel products has been done through austenitizing and quenching, often followed by a tempering treatment. These steels were typically rolled on a plate mill. The development of accelerated cooling equipment allowed quenching directly after plate rolling, simplifying the production process and improving the properties. Extensive literature and reviews on the topic can be found in references [1,2]. In more recent years, direct quenching processes have been developed for hot strip mills to produce ultra-high strength martensitic grades. The main advantage of the strip process compared to the plate process is the higher productivity for low thicknesses.

There exist two main classes of low carbon martensitic strip steels. The first is steels for wear applications. These steels mainly need to achieve a minimum hardness level, combined with a sufficient bendability and impact toughness. Typical applications include tipper trucks and mining equipment. The second class is martensitic strip steels developed for structural applications, e.g., trailers, chassis parts, and lifting equipment. In addition to a high strength, these steels also typically need to fulfill more stringent requirements for tensile elongation, impact toughness, and bendability, both for the as-produced sheets and for weldments.

This paper discusses the main metallurgical principles that need to be understood to develop low carbon martensitic strip steels via direct quenching. The weldability of the steels is also discussed in detail to give an indication of how these steels can be used in practice. Finally, the fatigue behavior in the deformed condition is described, relevant for evaluation of the fatigue performance of an actual structural component. The test that was developed shows the importance of an optimized metallurgical concept, e.g., the need for a low amount of inclusions.

2. Experimental Procedures

The steels used in this study were either lab or industrially produced casts, reheated to 1250 °C, followed by hot rolling with finish rolling temperatures (ranging from 860 to 1000 °C), and direct quenching to below 100 °C to ensure a martensitic microstructure. The chemical composition ranges of the three steels are given in Table 1.

Table 1. Chemical composition ranges of three direct quenched, ultra-high strength hot rolled steels. (All in wt %, balance: Fe + impurities). Measured via Optical Emission Spectroscopy, except C, N, and S via combustion.

Composition	C	Mn	Si	Cr + Mo + V	Cu + Ni	B	Ti + Nb + V
DQ1	<0.15	<2.0	<0.3	<0.50	<0.5	0.002	<0.100
DQ2	0.15–0.20	<2.0	<0.3	<0.50	<0.5	0.002	<0.100
DQ3	<0.15	<1.5	<0.3	<1.00	<0.5	0.002	<0.150

Tensile tests were performed on standard proportional tensile test specimens, in triplets, according to ISO 6892-1, using an Instron 4505 test rig (Norwood, MA, USA). Samples were taken along the rolling direction (RD). Triplets of Charpy impact tests were performed at 20 °C or −40 °C, using a Zwick pendulum impact tester PSW750 (Ulm, Germany). The tests were performed according to ISO 148-1. Samples were taken along the rolling direction (RD) with the V-notch along the normal direction (ND). Vickers hardness measurements (Future Tech, Kawasaki, Japan) were made using a 200 g weight, applied for 10 s. The load application was at a speed of 50 g/s. Optical microscopy (Olympus, Tokyo, Japan) after Béchet-Beaujard etching was used to reveal the prior austenite grain (PAG) structure. The prior austenite grain size (PAGS) was determined according to the linear intercept method (ASTM E112). Electron Back-Scatter Diffraction (EBSD) measurements were performed on a JEOL JSM-7001F FEG-SEM (Tokyo, Japan) at 20 kV equipped with a HKL Nordlys camera (Oxford, UK) for the detection of backscattered diffraction patterns. The angular resolution of the equipment is 2°. The step size used for the scan was 0.2 μm and the scanned area was 250×200 μm^2. The measured martensite orientation maps were used to reconstruct the prior austenite grain structure following the procedure described in [3]. Dilatometric experiments were performed in a Bahr DIL805 dilatometer (Hüllhorst, Germany), using cylindrical specimens with a diameter of 5 mm and a length of 10 mm. For cooling rates of above 25 °C/s, hollow 10 mm long cylinders with a diameter of 4 mm and a wall thickness of 1 mm have been used.

To investigate the weldability of the concepts, both welding and thermal simulation experiments were performed. For the welding, full penetration butt welds were made on a semi-V bevel using a Fronius TransPuls Synergic 3200 CMT gas metal arc welding (GMAW) power source (Wels, Austria), operated in standard mode (i.e., no pulsing of the welding current or periodical retraction of the filler wire). Welding was performed in the downhand (PA) position using a Panasonic TA-1800 six-axes robot-arm (Kadoma, Osaka, Japan). ESAB OK AristoRod 89 solid wire of 1 mm diameter was used as filler wire in combination with Air Liquide Atal 6 (82% argon-18% CO_2) M21 shielding gas. No pre- or post-heating was used, and the interpass temperature was kept below 100 °C. The critical welding parameters for calculation of the heat input (voltage, current, and travel speed) were measured using a TVC Arc Logger Ten (Great Yarmouth, UK). From the combination of the heat input, plate thickness, and the plate starting temperature, the cooling time from 800 to 500 °C—which is often seen as critical in welding—was calculated. Welding parameters were chosen to realize certain heat input levels and thus cooling times, rather than optimizing for weld shape.

In addition, samples of $11 \times 11 \times 70$ mm^3 were prepared and heat-treated using a Gleeble 1500 (Poestenkill, NY, USA), to simulate different parts of the heat-affected zone (HAZ), in particular the grain coarsened HAZ (GC HAZ), the inter-critical HAZ (IC HAZ), and the inter-critically reheated grain coarsened HAZ (IRGC HAZ). The difference between these regions is the peak temperature used (1350 and 775 °C for the grain coarsened and intercritical regimes, respectively); the last being a double-cycle treatment whereas the first two are single-cycle. The heating was performed at a rate of 100 °C/s up to 800 °C, then 300 °C/s to the peak temperature where a soaking time of 1 s was used. It was followed by cooling using different cooling times between 800 and 500 °C ($t8/5$). In the case of a dual cycle, the sample was cooled down to below 100 °C before the second cycle was started. The treated samples were then machined back to standard Charpy V notch (CVN) samples of $10 \times 10 \times 5$ mm^3 for CVN testing and per treatment condition, a sample was also used for hardness testing.

Force controlled fatigue tests were performed on specimens that were bent with a tool radius r of three-times the sheet thickness t. The set-up is shown in Figure 1. The fatigue experiments were carried out using six different axial load (stress) levels from 30 to 70 kN to produce the S–N curve. The fatigue loads were applied with a 0.1 loading ratio at a frequency of 2 Hz. At least two fatigue tests were performed at each load (stress) level to check the repeatability of results. For more information regarding the developed test method and the data analysis, readers are referred to [4].

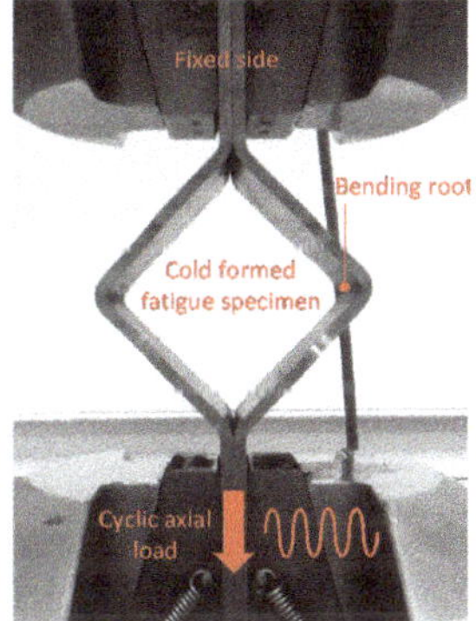

Figure 1. Fatigue test set-up of cold formed specimen after 90° bending with a bending ratio $r/t = 3$.

3. Metallurgical Concepts

3.1. Hot Strip Mill Parameters

The hot strip mill rolling schedule is the main parameter to optimize the strength/toughness performance and the effect of the finish rolling temperature (FRT) on the properties and microstructure is therefore discussed. Figure 2a–c shows EBSD scans of three plates of the DQ1 steel that were rolled on a lab hot rolling mill with different FRTs. An in-house developed routine [3] was applied to reconstruct the prior austenite structure, as shown in Figure 2d–f. The prior austenite grain size (PAGS), as determined through these reconstructed EBSD maps, showed no significant differences for the different conditions.

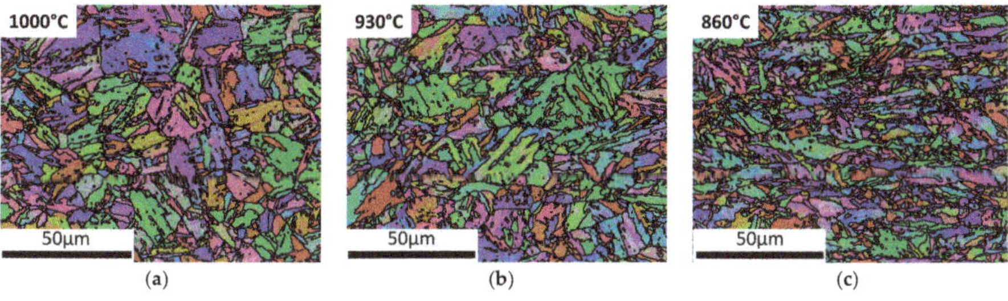

Figure 2. *Cont.*

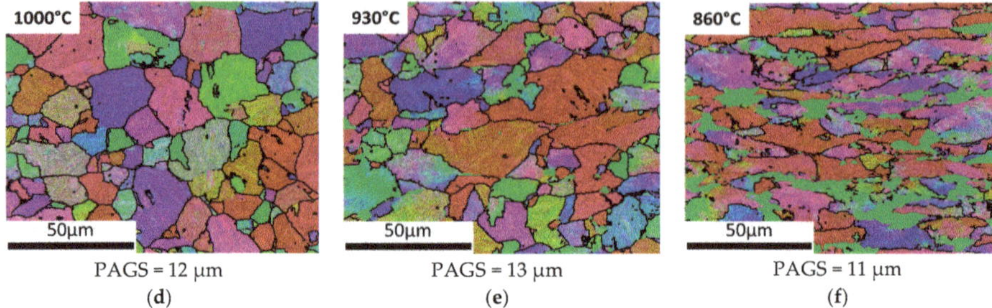

Figure 2. EBSD maps for three different finish rolling temperatures (FRTs) (DQ1 steel): originally measured martensite structure (**a**) FRT 1000 °C, (**b**) FRT 930 °C, (**c**) FRT 860 °C, and reconstructed prior austenite EBSD maps of the same measurements (**d**) FRT 1000 °C, (**e**) FRT 930 °C, (**f**) FRT 860 °C. (Inverse Pole Figure colouring, black lines: 15° boundaries) (Bright green areas in (**e**,**f**): unreconstructed areas).

When reducing the FRT, the shape of the prior austenite grains becomes more and more pancaked. The temperature of non-recrystallization (T_{nr}) was estimated to be around 930 °C [5]. Figure 3 shows that the best combination of tensile strength and Charpy impact toughness was obtained for a FRT of 860 °C, resulting in a PAGS aspect ratio of around 4 for the specific rolling schedule used. This positive effect of a rather moderate rolling reduction below the T_{nr} is in line with earlier findings in Mo-V-B alloyed low C steels [6]. Stronger effects on the strength-toughness have been reported for higher rolling reductions [7], although excessive austenite pancaking impairs the bendability of the steel [8]. The Charpy impact toughness is quite similar for all three conditions, which is a consequence of their very similar PAGS [5]. The strength, however, increased with reducing FRT. This can be attributed to a higher dislocation density in the martensite, which is apparent from the misorientation distributions as a higher fraction of low angle misorientations was found [9], as illustrated in Figure 3b.

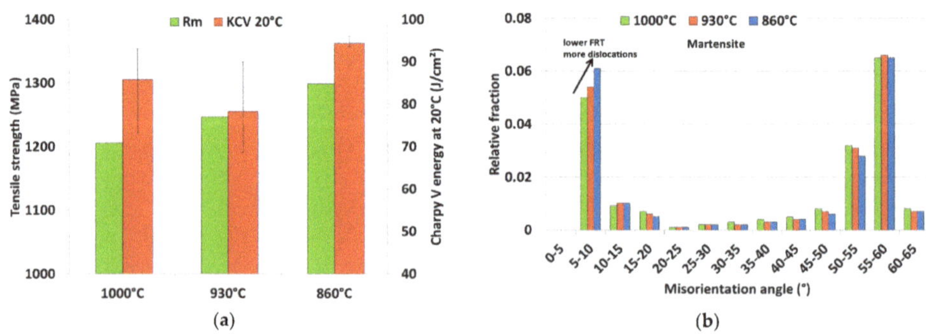

Figure 3. (**a**) Tensile strength and Charpy V impact energy at 20 °C, tested in rolling direction, (**b**) Misorientation angles, based on martensite EBSD map. (DQ1 steel).

3.2. Laminar Cooling Strategy

Since these steel grades are targeted to be martensitic directly after hot rolling, a high cooling rate is needed. The critical cooling rate needed varies with the chemical composition, as illustrated by the dilatometer curves in Figure 4. The lower alloyed DQ2 composition cooled at 10 °C/s showed significant amounts of ferrite and bainite. This is illustrated by the DQ2, 10 °C/s curve, which deviates from linearity above 600 °C, indicative of ferrite (α) formation. The transformation then gradually changes from ferrite to bainite (α_B), with the final fraction of austenite transforming to martensite (α_M)

at 425 °C. The DQ3 composition shows a much higher hardenability, without ferrite formation and with only a minor fraction of bainite formed during cooling at 10 °C/s, as shown by the deviation of linear behavior at 475 °C. Both steels, however, are fully martensitic for cooling rates of 20 °C/s (and higher), with the martensite start temperature M_s estimated to be around 415 °C for the DQ2 and around 445 °C for the DQ3 composition. Such a cooling rate is readily feasible with modern laminar cooling equipment on the run-out-table for typical strip thicknesses.

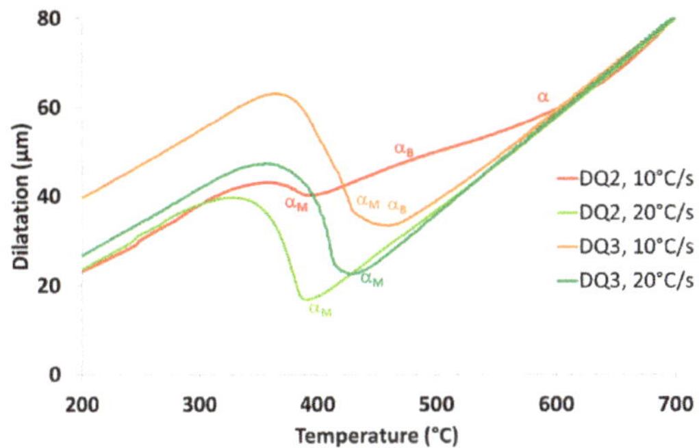

Figure 4. Dilatometry curves for DQ2 and DQ3 steels, cooling rates of 10 and 20 °C/s. (Thermal cycles: heating to 850 °C with 10 °C/s, holding time 300 s before final cooling).

3.3. Temper Annealing

A typical microstructure after temper annealing at 550 °C is shown in Figure 5a,b, showing fine martensite laths and temper carbides. Figure 6a–c compares the tensile properties and the impact toughness for the DQ1 and DQ3 compositions after temper treatments at different temperatures. The hot rolled processing and quenching was similar and the tempering time was kept constant as well. The DQ1 composition showed very strong softening with increasing tempering temperature. The DQ3 composition, however, maintains its high strength up to tempering temperatures of 600 °C. This is direct consequence of the presence of (Cr + Mo + V), which are known to suppress temper softening [10,11]. In addition to suppression of temper softening, the Charpy impact toughness of the DQ3 composition is also maintained at a high level. This is likely a consequence of the fact that the fine lath martensite structure is maintained up to high temperatures in the DQ3 steel, whereas the lath structure disappears in the DQ1 due to absence of (Cr + Mo + V) [12]. This is illustrated in Figure 5c.

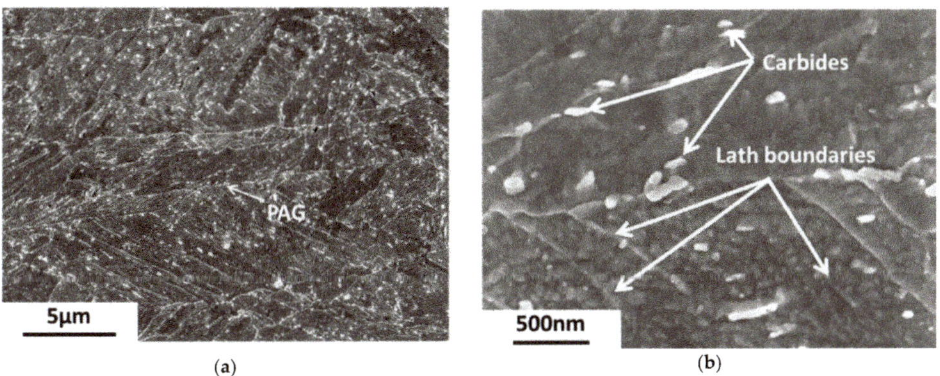

(a)

(b)

Figure 5. *Cont.*

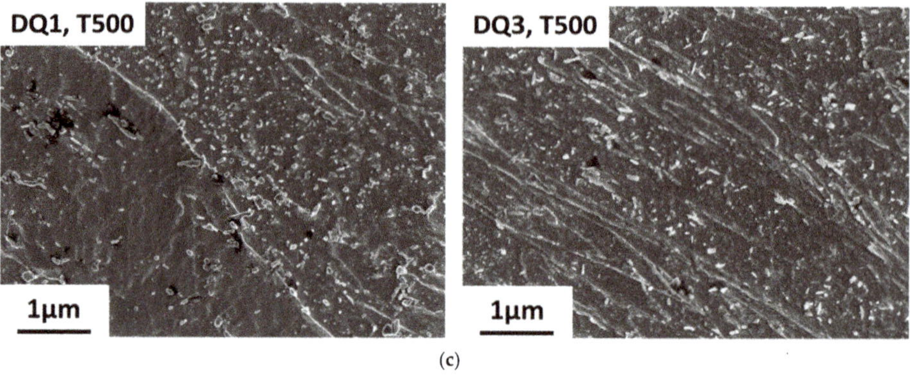

Figure 5. (**a**) Typical SEM micrograph of DQ3 steel (PAG = prior austenite grain), tempered at 550 °C (RD-ND section), (**b**) High magnification SEM micrograph showing temper carbides and residual martensitic lath structure of DQ3 steel, tempered at 550 °C, (**c**) Difference in remaining martensitic lath structure between DQ1 and DQ3 steel after tempering at 500 °C.

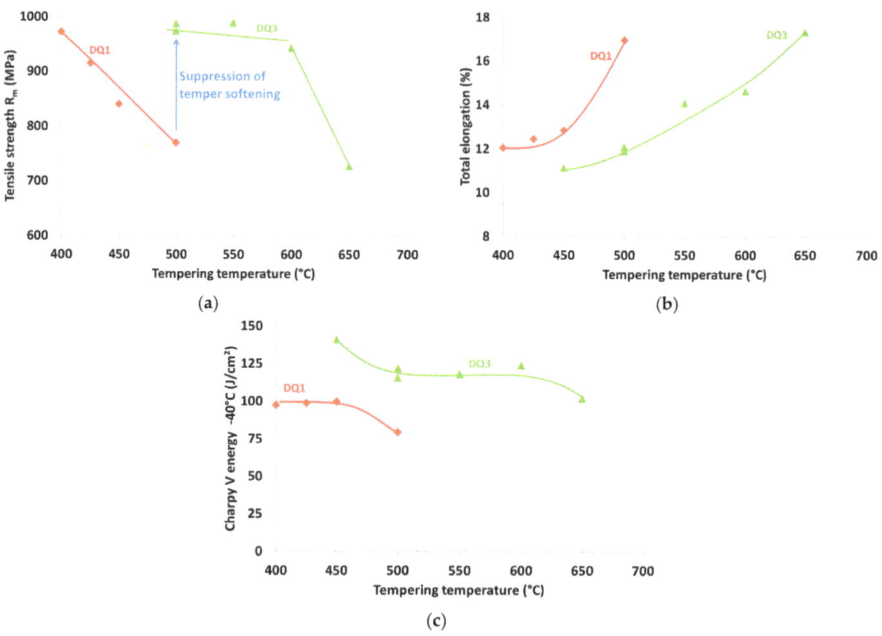

Figure 6. Comparison of mechanical properties of DQ1 and DQ3 steels for different temperatures (**a**) Tensile strength, (**b**) Total elongation (proportional test specimen), (**c**) Charpy V impact energy at −40 °C.

4. In-Use Properties

4.1. Welding

In terms of weldability of martensitic structural steels, the two main properties that need to be maintained are the hardness (and thus strength) and the impact toughness. For the hardness, the same alloying additions that suppress temper softening (Cr, Mo, V) also help maintain the hardness in the HAZ after welding [13]. This is illustrated by the hardness profiles in Figure 7 of similar weldments made with gas metal arc welding (GMAW) of the DQ1 and DQ3 alloys. Figure 7b clearly demonstrates the better softening resistance of the more highly alloyed DQ3. Nonetheless, even in the case of the DQ3, too high a heat input will still result in unacceptable softening in the grain coarsened heat affected zone (GCHAZ), as illustrated in Figure 8.

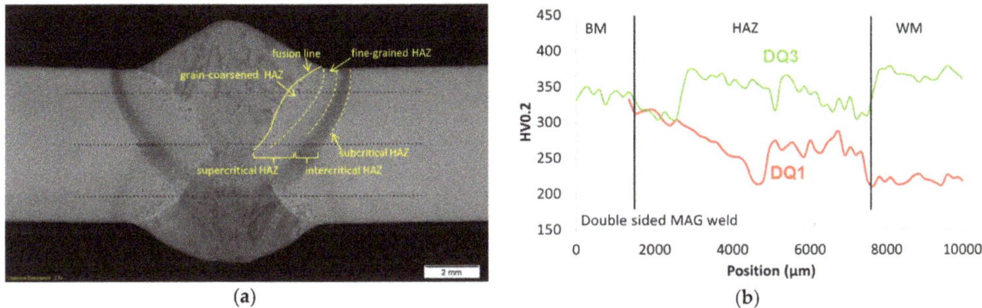

(**a**)   (**b**)

Figure 7. (**a**) Lab produced double-sided GMAW weld (Note. weld made with a specified heat input; not optimized for weld shape), (**b**) Hardness profile across the heat-affected zone (HAZ) of similar double-sided GMAW welds for the DQ1 and DQ3 steels. (Thickness: 6 mm, $t8/5 = 7$ s).

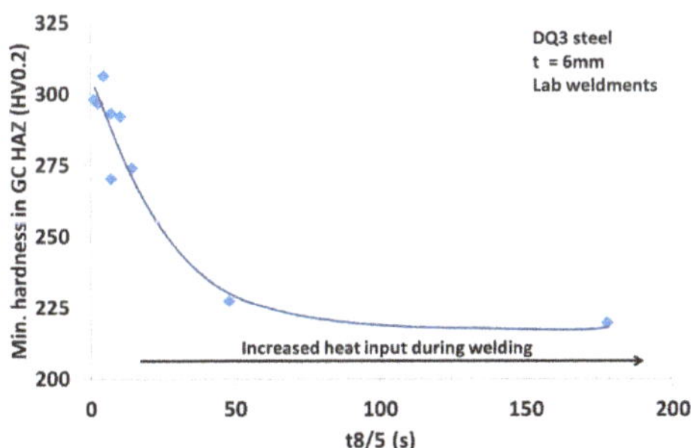

Figure 8. Minimum hardness measured in the GC HAZ for lab welds of DQ3 with different heat inputs ($t8/5$ = cooling time from 800 to 500 °C).

To identify the maximum allowable heat input and corresponding 800 to 500 °C ($t8/5$) cooling times for the DQ3, annealing simulations of the Grain Coarsened (GC HAZ), intercritical (IC HAZ), and intercritically reheated grain coarsened (IRGC HAZ) HAZ were made using a Gleeble thermo-mechanical simulator. Charpy testing results of the Gleeble samples in Figure 9a show that a minimum Charpy impact energy of 34 J/cm^2 can be expected for all parts of the HAZ tested for a $t8/5$ range from at least 6 to 15 s. For the IC HAZ, some individual values are below the target, but the average values are still above the minimum required and the individual values above 70% of that. A welding continuous cooling diagram (CCT) of the GC HAZ was also constructed using dilatometry to establish the allowable cooling time range to maintain the hardness. The CCT diagram simulation (Figure 9b) of the Grain Coarsened HAZ (GC HAZ) shows that softening starts to take place for $t8/5$ cooling times above 10–15 s. The softening observed is a consequence of the formation of bainite instead of martensite. Excessive softening occurs only for cooling times above 25 s, which is in line with the results on the lab welds in Figure 8.

Based on the results obtained, the allowable $t8/5$ cooling times for welding of the DQ3 are set between 6 s and at least 15 s.

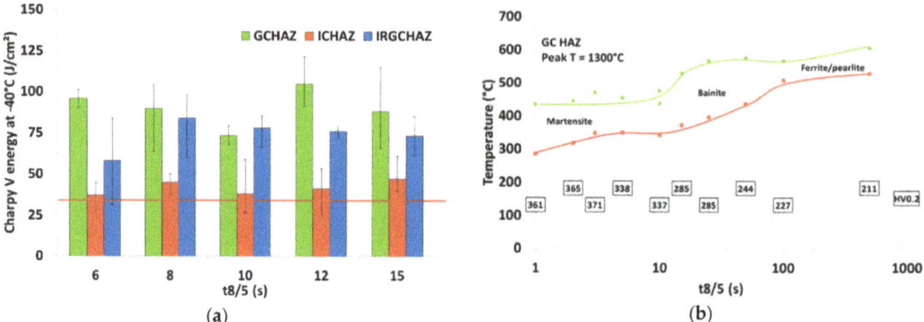

Figure 9. Welding simulation data for DQ3 steel: (**a**) Charpy V impact energy at $-40\,°C$ on Gleeble annealed samples with different $t8/5$ (GC HAZ: peak temperature 1350 °C, ICHAZ: peak temperature 775 °C, IRGZ HAZ: dual cycle: $T_{p1} = 1350\,°C$; $T_{p2} = 775\,°C$), (**b**) Welding CCT diagram for the CG HAZ (Thermal cycle: Heating at 100 °C/s up to 800 °C, then 300 °C/s to peak temperature 1300 °C, followed by cooling using different cooling times between 800 and 500 °C ($t8/5$)).

4.2. Fatigue Performance of Deformed Components

In order to obtain good behavior of a deformed component in fatigue, it is crucial that the damage to the microstructure after forming—typically bending—remains limited. In Figure 10a,b, the result after 90° bending using a radius of 1.8 times the thickness is shown for two DQ3 materials with different nitrogen (N) levels. The cracked sample in Figure 10a has a high N level, whereas the N level in the intact one in Figure 10b has a low N level. The micrograph in Figure 10c shows that significant microstructural damage occurs around the coarse TiN particles during deformation. Note that the chemical composition of this specific particle was not determined, but it has been assumed to be TiN based on its shape and size.

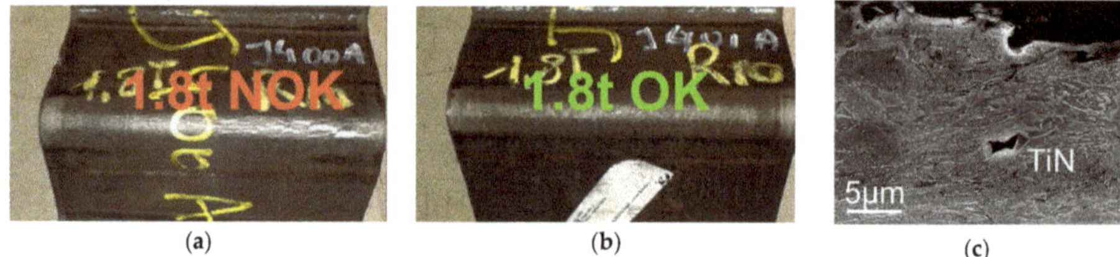

Figure 10. Effect of N on the bending behavior, 90° V-bending, $r/t = 1.8$ (**a**) DQ3 steel, high N; (**b**) DQ3 steel, low N; (**c**) Damage around a coarse TiN particle in DQ1 steel.

There are many steel components for different applications that are subjected to both bending and fatigue loading. Combining these two—i.e., cold forming followed by fatigue loading—causes micro-crack initiation at the bending root and fatigue propagation of the cracks up to final rupture of material. For this purpose, a new test sample was designed to take into consideration the effect of cold forming before the fatigue load is applied. For more information regarding the developed test method, readers are referred to [4].

The fatigue tests of the cold formed fatigue specimens were performed on a hydraulic testing machine as shown in Figure 11a. Figure 11b shows the maximum stress value at the bending root versus the number of cycles to final failure of the cold formed fatigue specimen. The maximum stress at the bending root, as depicted in Figure 11a, was calculated using a finite element model (see [4]).

The observed results show the effect of the N level on the fatigue response of cold formed fatigue samples. By reducing the N level, the fatigue cycles to failure increase considerably, as shown in Figure 11. It is known that large, non-metallic particles reduce the fatigue performance of metals by microcrack formation at the matrix-particle interface [14]. This effect is likely further strengthened in the case of pre-existing damage as a consequence of deformation (cf. Figure 10c). A lower N content results in fewer and smaller TiN particles, as shown in Figure 12. This will result in less damage during the bending process for the low N steel, which in turn improves the fatigue performance of cold formed components. This effect is more pronounced at lower stress levels.

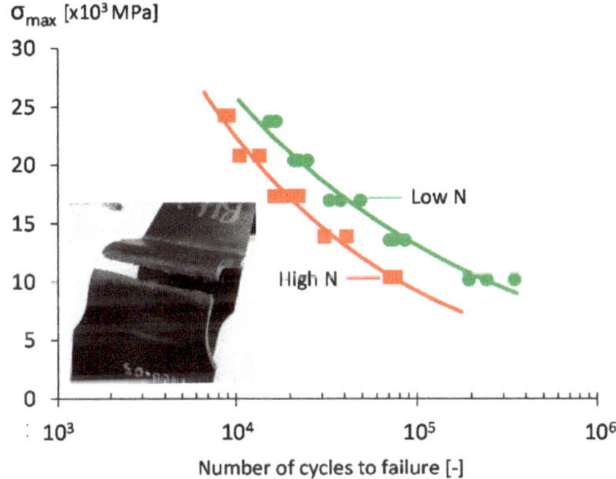

Figure 11. S–N curve obtained from cold formed specimen. The maximum stress at the bending root was calculated using a finite element model.

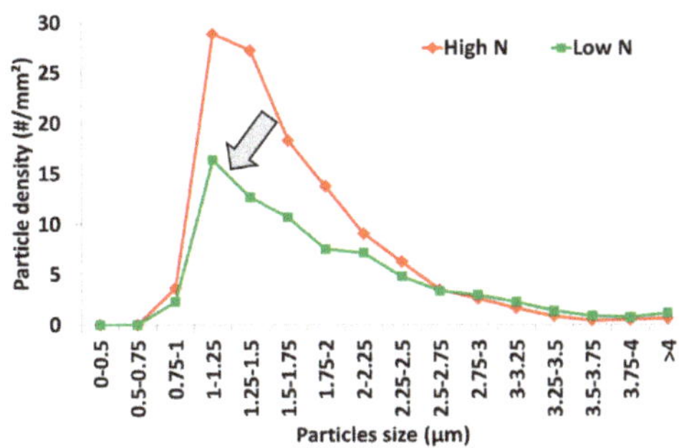

Figure 12. TiN particle size distribution for the high N and low N steel (SEM analysis: chemical composition measured by EDX, size measured by automated procedure based on image contrast).

5. Summary and Conclusions

The main metallurgical parameters for the development of ultra-high strength martensitic strip steels have been discussed. Furthermore, the welding and fatigue behaviors of these steels have been illustrated.

The main parameter to optimize the strength/toughness balance is the finish rolling temperature after hot rolling. The FRT should be below the temperature of non-recrystallization, resulting in an optimum prior austenite grain size aspect ratio around 4. An additional tempering treatment in the range of 500 to 600 °C can be applied to further improve the toughness while maintaining the strength. The latter requires the addition of alloying additions that suppress temper softening, like Cr, Mo, and V. These same alloying additions also help to suppress softening during welding.

A specific test set-up was designed to investigate the fatigue behavior of a deformed component. It was shown that the fatigue behavior can be significantly improved if the N level is kept low. This is due to lower microstructural damage around TiN particles after forming, but likely also due to an intrinsic better fatigue response of the undeformed material.

Author Contributions: Lieven Bracke: contribution to all sections, related to metallurgical aspects; Dorien De Knijf, Eva Diaz Gonzalez: processing and metallurgy; Christoph Gerritsen: section on welding; Reza Hojjati Talemi: section on fatigue.

Conflicts of Interest: The authors declare no conflict of interest.

References

1. Taylor, K.A.; Thompson, S.W.; Fletcher, F.B. (Eds.) Physical Metallurgy of Direct-Quenched Steels. In Proceedings of the Minerals, Metals, and Materials Society/American Society for Metals (TMS/ASM) Materials Week 92 Conference, Chicago, IL, USA, 2–4 November 1992.
2. Ouchi, C. Development of Steel Plates by Intensive Use of TMCP and Direct Quenching Processes. *ISIJ Int.* **2001**, *41*, 542–553. [CrossRef]
3. Bernier, N.; Bracke, L.; Malet, L.; Godet, S. An alternative to the crystallographic reconstruction of austenite in steels. *Mater. Charact.* **2014**, *89*, 23–32. [CrossRef]
4. Talemi, R.H.; Chhith, S.; De Waele, W. Experimental and numerical study on effect of forming process on low cycle fatigue behavior of high strength steel. *Fatigue Fract. Eng. Mater. Struct.* **2017**. [CrossRef]
5. Bracke, L.; Xu, W.; Waterschoot, T. Effect of finish rolling temperature on direct quenched low alloy martensite properties. *Mater. Today Proc.* **2015**, *2*, S659–S662. [CrossRef]
6. Taylor, K.A.; Hansen, S.S. Effects of Vanadium and Processing Parameters on the Structures and Properties of a Direct-Quenched Low-Carbon Mo-B Steel. *Metall. Mater. Trans. A* **1991**, *22*, 2359–2374. [CrossRef]
7. Kaijalainen, A.J.; Suikkanen, P.P.; Limnell, T.J.; Karjalainen, L.P.; Kömi, J.I.; Porter, D.A. Effect of austenite grain structure on the strength and toughness of direct-quenched martensite. *J. Alloys Compd.* **2013**, *577*, S642–S648. [CrossRef]
8. Kaijalainen, A.J.; Suikkanen, P.; Karjalainen, L.P.; Jonas, J.J. Effect of Austenite Pancaking on the Microtexture, Texture, and Bendability of an Ultrahigh-Strength Strip Steel. *Metall. Mater. Trans. A* **2014**, *45*, 1273–1283. [CrossRef]
9. Zhao, Y.; Shi, J.; Cao, W.; Wang, M.; Xie, G. Effect of direct quenching on microstructure and mechanical properties of medium-carbon Nb-bearing steel. *Univ.-Sci. A Appl. Phys. Eng.* **2010**, *11*, 776–781. [CrossRef]
10. Klein, M.; Rauch, R.; Spindler, H.; Stiaszny, P. Ultra high strength steels produced by thermomechanical hot rolling: Advanced properties and applications. In Proceedings of the 3rd International Conference on Steels in Cars and Trucks (SCT) 2011, Salzburg, Austria, 5–9 June 2011.
11. Grange, R.A.; Hribal, C.R.; Porter, L.F. Hardness of Tempered Martensite in Carbon and Low-Alloy Steels. *Metall. Mater. Trans. A* **1977**, *8*, 1775–1785. [CrossRef]
12. Caron, R.N.; Krauss, G. The Tempering of Fe-C Lath Martensite. *Metall. Mater. Trans. B* **1972**, *3*, 2381–2389. [CrossRef]
13. Humber, G.; Klein, M.; Spindler, H.; Ernst, W. Properties and metallurgical aspects of thin wear resistant steel sheets of hardness 400/450HB produced in a hot strip mill. In Proceedings of the 4th International Conference on Steels in Cars and Trucks (SCT) 2014, Braunschweig, Germany, 15–19 June 2014.
14. ASM International. Fatigue and Fracture. In *ASM Handbook*; ASM International: Geauga County, OH, USA, 1996; Volume 19, p. 155.

 *metals*

Article

Effect of Austempering Time on the Microstructure and Carbon Partitioning of Ultrahigh Strength Steel 56NiCrMoV7

Quanshun Luo *, Matthew Kitchen and Shahriar Abubakri

Materials and Engineering Research Institute, Sheffield Hallam University, Howard Street, Sheffield S1 1WB, UK; acesmk@exchange.shu.ac.uk (M.K.); acessa1@exchange.shu.ac.uk (S.A.)
* Correspondence: q.luo@shu.ac.uk; Tel.: +44-114-2253649

Received: 23 May 2017; Accepted: 3 July 2017; Published: 7 July 2017

Abstract: Ultrahigh strength steel 56NiCrMoV7 was austempered at 270 °C for different durations in order to investigate the microstructure evolution, carbon partitioning behaviour and hardness property. Detailed microstructure has been characterised using optical microscopy and field emission gun scanning electron microscopy. A newly developed X-ray diffraction method has been employed to dissolve the bainitic/martensitic ferrite phase as two sub-phases of different tetragonal ratios, which provides quantitative analyses of the carbon partitioning between the resultant ferrites and the retained austenite. The results show that, a short-term austempering treatment was in the incubation period of the bainite transformation, which resulted in maximum hardness being equivalent to the oil-quenching treatment. The associated microstructure comprises fine carbide-free martensitic and bainitic ferrites of supersaturated carbon contents as well as carbon-rich retained austenite. In particular, the short-term austempering treatment helped prevent the formation of lengthy martensitic laths as those being found in the microstructure of oil-quenched sample. When the austempering time was increased from 20 to 80 min, progressive decrease of the hardness was associated with the evolution of the microstructure, including progressive coarsening of bainitic ferrite, carbide precipitating inside high-carbon bainitic ferrite and its subsequent decarbonisation.

Keywords: ultrahigh strength steel; austempering; carbon partitioning; carbide precipitation; bainitic/martensitic ferrite

---

1. Introduction

Low alloy ultrahigh strength steels are key structural materials for load-bearing components of large vehicles and aircrafts [1–3]. In addition to the conventional strengthening treatments of quenching and tempering (QT) and bainitic salt bathing process, a new treatment has been developed by lowering the salt bath temperature to a level close to, or even lower than, the Ms (martensite starting) temperature [4–10]. The resultant microstructure comprises extremely fine ferrite laths and inter-lath filmy austenite, both having super-saturated carbon contents because of the prohibition of carbide precipitation. Such ferrite based microstructure enables ultrahigh strength properties, e.g., 2.5 GPa of compression strength and 2.3 GPa of ultimate tensile strength. However, such a process needs a very long salt bath time, and concern also arises from the low toughness property [5,6,11].

When an austempering treatment is applied, wherein the austenised steel is firstly soaked for a short period in a salt bath of lower-bainite transformation and is subsequently quenched to room temperature, the resultant microstructure is a mixture of bainite, martensite and retained austenite [3,12–17]. The new heat treatments were reported to bring about higher hardness and strength properties than the QT treatments, while the strengthening mechanisms have not been fully understood up to date. Tomita and co-workers attributed the increased strength properties to several factors, including martensite grain refining, due to partitioning of austenite by bainitic ferrite prior to martensite transformation, the predominant role of the major martensite plates, and strain-hardening of the bainitic ferrite, induced by the martensite transformation [18]. These explanations were also adopted by other researchers [13,16–18]. However, the contribution of supersaturated carbon to hardening was excluded in this model.

In recent years, increasing attention has been paid on the role of carbon in steel strengthening. Supersaturated carbon is highly involved in several strengthening mechanisms of ultrahigh strength steels, regardless of the transformation kinetics [19–21]. In the transformation from austenite to martensite or bainite, carbon is known to diffuse from newly formed martensite or bainite to the parent austenite. This phenomenon, known as carbon partitioning, has been utilised to develop a novel toughening treatment, i.e., the so-called quenching and partitioning (Q–P) treatment [22–25]. Although carbon partitioning as a general phenomenon was noticed a long time ago [2,26], it is in the Q–P process that the phenomenon was used the first time to control the transformed microstructure [22].

In a recent project, we worked on austempering treatments of a Cr–Ni–Mo–V alloyed spring steel, which led to superior mechanical properties, including ultimate tensile strength of 2100 MPa, yielding strength of 1800 MPa and elongation of 8–10%, along with V-notched Charpy impact toughness of 9–12 J/cm^2 [27,28]. This paper reports the microstructure characterisations of a series of austempered samples, with a focus on the carbon partitioning behaviour at various isothermal soaking times.

2. Experimental

The sample material was vacuum-re-melted steel and provided as a 50 mm diameter hot-rolled bar with nominal chemical compositions (in wt %) of C 0.55, Si 0.30, Mn 0.76, P 0.014, S 0.0037, Ni 1.69, Cr 1.05, Mo 0.50, V 0.08, and Fe in balance. The A_{c3}, A_{c1} and M_s temperatures of the steel are estimated to be 780, 715, and 240 °C respectively, as provided by the steel supplier (Böhler Tool Steel & High Speed Steel, Oldbury, UK). The bar was cut as block samples of 20 mm × 10 mm × 8 mm in size for austempering heat treatments. In the heat treatments, the samples were heated in the first salt bath to soak at 850 °C for 30 min. Then they were moved immediately to the second salt bath to soak at 270 °C (i.e., 30 °C above the M_s, at the range of lower bainite transformation) for a selected time before air-cooling to room temperature. The austempering soaking times were selected to be 5, 10, 20, 40 and 80 min.

The austempered samples were mounted, and manually ground to remove a surface layer of about 0.3 mm in thickness using coarse SiC abrasive paper (grade 120 #). Then fine grinding was applied using SiC abrasive papers of grades 240 and 600 # for hardness testing. Vickers hardness tests were undertaken at an indenting load of 30 kg. Five indents were made on each sample to calculate the average hardness and the standard deviation. Then following metallographic grinding and polishing, the samples were etched using a 2%-nital etchant before microstructure characterisations using optical microscopy (OPM, Olympus BX51, Tokyo, Japan), scanning electron microscopy (SEM) and X-ray diffraction (XRD).

A FEI Nova200 field emission gun (FEG, Eindhoven, The Netherlands) SEM was employed, where the field emission gun provides a spatial resolution of 1.8 nm, being sufficient to resolve nano-scale structural features. A X'Pert X-ray diffractometer (PANalytical, Almelo, The Netherlands) with Cu-K$_\alpha$ radiation (wavelength λ = 0.15406 nm) was employed in the study. On each sample, five diffraction peaks were acquired under the θ–2θ (Bragg-Brentano) mode, including the austenite peaks (200)$_\gamma$, (220)$_\gamma$ and (311)$_\gamma$, and the ferrite peaks (200)$_\alpha$ and (211)$_\alpha$. The obtained diffractions were measured using a self-developed Gaussian peak-fitting technique to determine the peak position (2θ), net intensity (I) and broadening width (β) (FWHM, the full-width at half maximum) [27,28]. These parameters were then used to calculate the volume fraction (V_γ) and carbon content (C_γ) of retained austenite and the micro-strains (ε) of both the martensitic and bainitic ferrite and the retained austenite. The equations for the calculations were adapted from literature [29,30].

$$V_\gamma = I_\gamma \times (I_\gamma + G \times I_\alpha)^{-1} \tag{1}$$

$$C_\gamma = (a_\gamma - 0.3555)/0.0044 \tag{2}$$

$$\varepsilon = 0.25 \times \beta \times \cos\theta \times \sin\theta^{-1} \tag{3}$$

In Equation (2), a_γ stands for the austenite lattice parameter, which is an average of three lattice parameter values derived from the $(200)_\gamma$, $(220)_\gamma$ and $(311)_\gamma$ measurements. The $(200)_\alpha$ diffraction peak was numerically processed more extensively using a newly developed quantitative method. Details of the new method have been described elsewhere [31]. The new analytical method considers martensite in a hardened medium-carbon steel as a mixture of lath- and plate-martensites, whereas the two martensite sub-phases are known to have different tetragonal ratios because of their different carbon contents.

3. Results

3.1. Hardness Property of Austempered Samples

Figure 1a shows the effect of austempering time on the hardness property. When a short-time austempering was applied, the steel remained a high level of hardness, being even slightly higher than the as-quenched sample (HV 679). The maximum hardness of HV 716 was measured in the sample that was austempered for 10 min; then, further increased austempering time from 10 to 40 min resulted in remarkably decreased hardness. Following that, the hardness stabilised in that level when the austempering time was extended to 80 min. These results suggest that a short period of isothermal treatment did not soften the steel, while the steel became substantially softer when the treatment was longer.

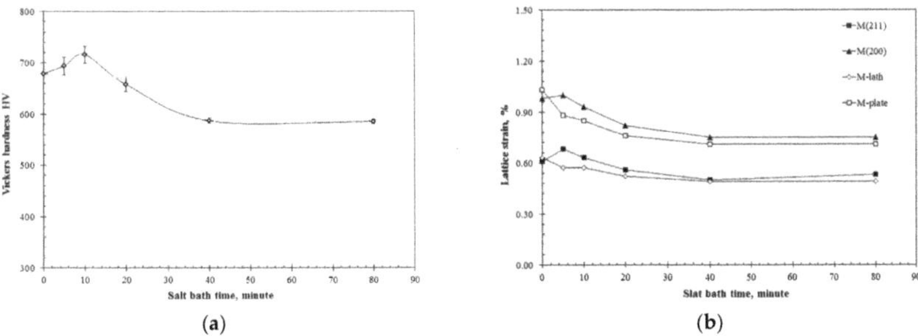

Figure 1. Effect of austempering time on (**a**) the hardness property of the sample steel and (**b**) the lattice strains of the martensitic and bainitic ferrites.

Figure 1b illustrates the variation of overall lattice strains of the martensitic and bainitic ferrites as determined by XRD peak broadening analyses. Firstly, both the overall diffraction peaks $(200)_\alpha$ and $(211)_\alpha$ were analysed to calculate the lattice strains along the <200> and <211> directions. Then by Gaussian separation of the $(200)_\alpha$ peak as sub-peaks, the martensitic or bainitic ferrite was considered as two sub-phases having different tetragonal ratios, namely lath martensite (M_{lath}) and plate martensite (M_{plate}) [31]. Consequently the lattice strains of the M_{lath} and M_{plate} sub-phases can be estimated. The analyses reveal correlation between the hardness property and the lattice strains of the martensitic and bainitic ferrite, which is consistent with the knowledge that martensitic transformation is the predominant hardening mechanism. The strains decrease gradually with increasing austempering time. The calculated strain, consistent with the hardness property, can be attributed to both the recovery of defects and the ferrite decarbonisation as a result of carbide precipitation. Moreover, in the ferritic sub-phases, the higher micro-strain values of the M_{plate} reveal its substantially higher lattice distortion than the M_{lath}.

3.2. Microstructure Observed in Optical Microscopy

In OPM observations, bainitic ferrite was easily recognised as black acicular grains. The OPM images displayed in Figure 2 show variation of the martensitic and bainitic microstructure with increasing austempering time. The oil-quenched sample was known to have predominantly martensitic structure, whereas it also shows some black-etched acicular grains similar to bainite. The black contrast in these acicular grains suggests fine carbide precipitates, which has been confirmed in subsequent SEM observations at higher magnification, to be discussed later in Figure 3. The sample austempered in 5 min exhibits mostly martensitic microstructure, comparable to the oil-quenched sample, as shown in Figure 2a,b. When the austempering time was increased to 10 and 20 min, the steel shows a mixed microstructure of bainite and martensite where the amount of bainite increases with the austempering time, (Figure 2c,d). In Figure 2c,d, the bainite regions exhibit black contrast because of the preferential etching of the bainite grains as compared to the martensite. The bainite and martensite regions are labelled "B" and "M", respectively.

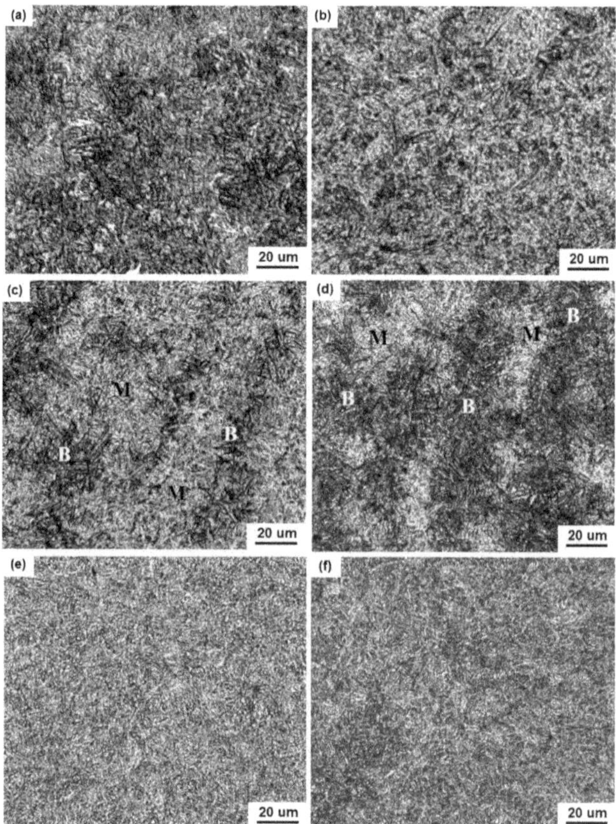

Figure 2. Optical micrographs of samples after 270 °C austempering treatment for various durations: (**a**) oil-quenched; (**b**) 5 min; (**c**) 10 min; (**d**) 20 min; (**e**) 40 min; and (**f**) 80 min.

The samples being austempered by 40 and 80 min are similar to each other and exhibit a relatively homogeneous contrast, because that bainitic transformation had taken place in the whole volume, as shown in Figure 2e,f. The change from heterogeneous to homogeneous microstructure suggests that the bainite transformation was completed in an intermediate time between 20 and 40 min, whereas further soaking after the completion did not give rise to noticeable change in the optical microstructure.

The OPM observations suggest that, bainite transformation did not take place in the initial short period, indicating the existence of a bainitic incubation period, as suggested by other researchers [32–34]. In the current work, noticeable bainite transformation took place after an incubation period of less than 10 min and, after that, significant austenite to bainite transformation occurred with increasing soaking time.

In addition, the OPM observations also suggest heterogeneous distribution of the bainitic structure. In Figure 2c,d, the bainite-rich regions exhibits black contrast as compared to the martensite-rich regions, in which initial bainite formed preferentially in the dendritic stems where C–Cr–Mo contents are lower than the inter-dendritic areas. The structural heterogeneity arose from the dendritic segregation of the as-cast steel. Such segregation was developed in the casting and was retained even after the hot rolling. Similar structural heterogeneity was also reported in other low alloy steels [35].

3.3. Microstructure Observed in Scanning Electron Microscopy

In Figure 3a,b, the microstructures of the 5 min austempered and oil-quenched samples are compared to each other at low magnification imaging, both showing as a mixture of lath- and plate-shape martensites. In drawing comparisons between Figure 3a,b, the martensites of austempered steel are more uniform and also slightly smaller in grain size. The comparison may reveal grain refining by means of the short austempering pre-treatment. Previously, researchers have attributed the grain refining to the increased ferrite nucleation sites [32–35]. They described such short-range motion of carbon atoms in the bainite incubation period as spinodal decomposition, which resulted in nano-scale heterogeneous distribution of carbon atoms to fertilise nucleation of ferrite in the carbon-depletion domains.

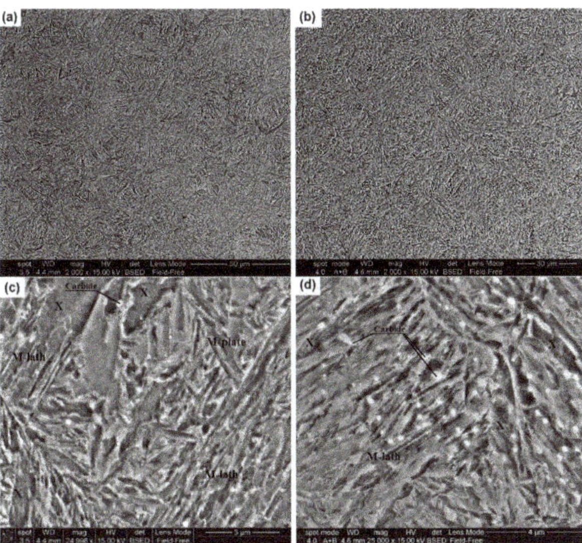

Figure 3. Field emission gun scanning electron microscopy (FEG-SEM) images of (**a**,**c**) the oil-quenched sample and (**b**,**d**) the 5 min austempered sample.

Details of the microstructure constituents are presented at high magnification in Figure 3c,d, in which the lath- and plate-shape martensites are labelled. Observations at higher magnifications showed that, some coarse martensite grains contain extremely fine particles, indicative of carbide precipitation, e.g., in the grains labelled "X". Such precipitation is more pronounced in the oil-quenched sample, which is shown in Figure 3c. On the other hand, most of the martensite grains of the austempered sample are free from carbide precipitation, as shown in Figure 3d.

In addition, small quantities of un-dissolved carbide particles can also be observed. At high magnification, these particles have a round shape and disperse heterogeneously in the matrix, as labelled "Carbides" in Figure 3c,d. The presence of un-dissolved carbides was attributed to insufficient decomposition of the cementite grains in the austenisation stage. In the heat treatment, the samples were heated to 850 °C and kept isothermally at that temperature for 30 min. The holding time was not long enough to have all the cementite carbide grains dissolve in the austenite matrix.

In a set of low-magnification SEM images, Figure 4 illustrates the bainitic grain growth with increasing austempering time. In Figure 4a, the 10 min treated sample exhibits a heterogeneous distribution of lath- and plate-shape martensite or bainite. The laths are fine and narrow, located mostly in the regions containing un-dissolved nodular carbide particles. In contrast, the rest regions show relatively coarse ferritic leaves, (the lightly etched, labelled "B"). These leaves are believed to be carbide-free bainitic ferrite by comparing them to the grain coarsening and carbide precipitation in similar regions. Nevertheless, the sample still shows fine grain size. Significant grain coarsening happened when the austempering time was increased to 20 and 40 min, as shown in Figure 4b,c. In Figure 4b, the less-etched blocks are mixtures of acicular martensite or bainite and blocky retained austenite (labelled "A"). A few small needle-like ferrite grains (labelled with an arrow) can be found inside a blocky austenite grain, implying fine martensite plates growing inside partitioned austenite, which is consistent with the grain refining mechanism proposed by Tomita [3]. In Figure 4c, bainitic ferrite becomes the dominant structural component in the 40 min treated sample and the inter-granular austenite has been much less. When the isothermal soaking time was increased to 80 min in Figure 4d, the whole image is almost full of bainitic ferrite grains, except for some very narrow inter-lath filmy austenite, as well as un-dissolved carbide particles.

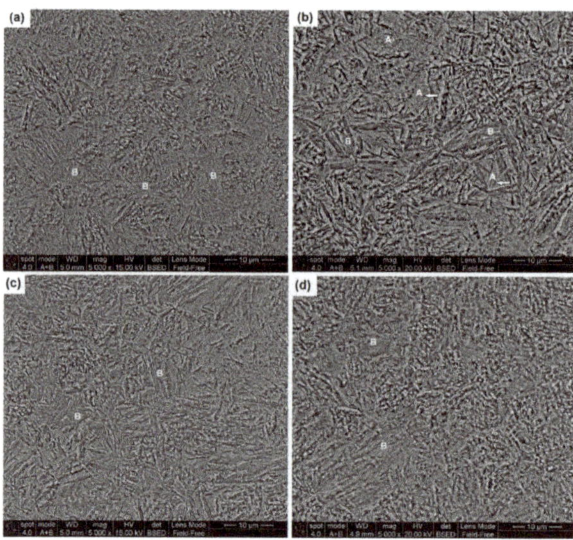

Figure 4. Low-magnification FEG-SEM images of samples austempered for different durations: (**a**) 10 min; (**b**) 20 min; (**c**) 40 min; and (**d**) 80 min. Note the growth of bainitic ferrite.

Figure 5 shows evolution of the microstructure at high magnification, wherein one can see the growth of carbide precipitates inside the bainitic ferrite grains. In Figure 5a,b, the 10 and 20 min treated samples show a mixture of bainite (labelled "B", containing carbide precipitates) and martensite (labelled "M", precipitate-free). In Figure 5c, some bainite grains contain carbide precipitates which are as fine as those shown in Figure 5b, whereas others show coarse precipitates. In Figure 5d, the microstructure shows significantly coarsened precipitates. It is believed that, the carbide coarsening followed the diffusion kinetics, i.e., taking place by absorbing carbon atoms from the ferrite matrix. As a result, the carbon content of the ferrite should be gradually lower, leading to decreased micro-straining, as shown in Figure 1b, and decreased lattice tetragoneity. This has been confirmed by the XRD analysis as described below.

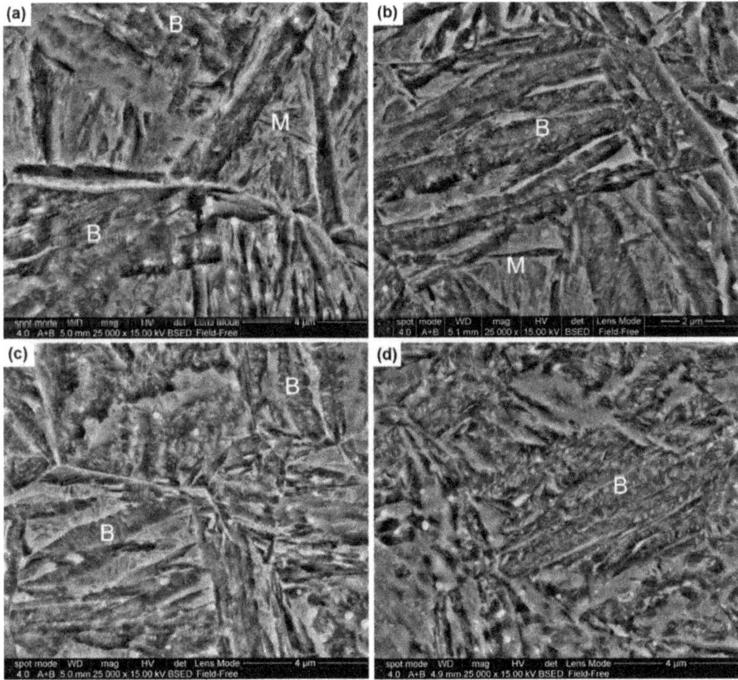

Figure 5. High-magnification FEG-SEM images of samples austempered for different durations: (**a**) 10 min; (**b**) 20 min; (**c**) 40 min; and (**d**) 80 min. Note the growth of bainitic ferrite.

3.4. Results of X-ray Diffraction Analyses

Figure 6 shows the collected diffraction peaks of both the austempered and oil-quenched samples. A quick comparative view reveals that the oil-quenched sample exhibits the maximum peak broadening, whereas the peaks of the austempered samples become increasingly sharper with increasing austempering time, indicative of the variation of lattice distortion. Meanwhile, retained austenite peaks are visible in most cases. However, the as-quenched sample does not present the maximum intensities in the austenite peaks, which would suggest increased austenite following the austempering treatments.

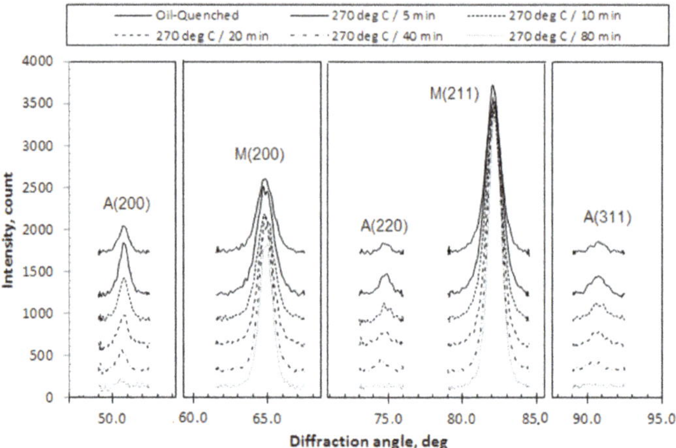

Figure 6. X-ray diffraction (XRD) peaks of the retained austenite and martensitic/bainitic ferrite phases.

Valuable results on the microstructure characteristics have been obtained from the quantitative analyses of these diffraction patterns, including the volume fraction, carbon content and micro-strain of retained austenite and two types of ferritic sub-phases. Detailed consideration of the ferritic sub-phases can be found in a recent publication [31], namely, that the mixture of martensite and bainite structure have been treated as two martensitic sub-phases (M_{plate} and M_{lath}) of different tetragonal ratios; although, a limitation of the method was that it was not capable of separating the austemperin-formed bainite and the quenching-formed martensites. Figure 7 shows an example of separation of the ferrite diffraction peak $(200)_\alpha$ as four sub-peaks using the Gaussian peak-fitting technique, where peaks 1 and 2 represent the M_{plate} diffractions and peaks 3 and 4 represent the M_{lath} diffractions. Then the integrated intensity, peak position and FWHM value of each sub-peak can be measured to calculate the crystalline characteristics. The results are shown in Figure 8.

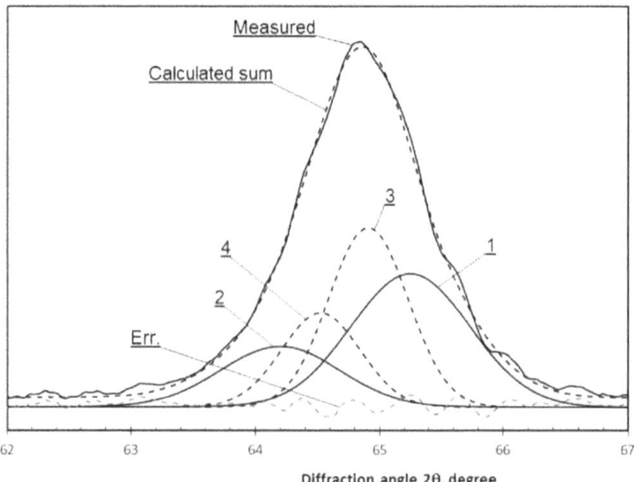

Figure 7. Multiple Gaussian peaking fitting to separate the $(200)_\alpha$ peak as four sub-peaks of the M_{plate} and M_{lath} sub-phases as labelled: 1—(200) of M_{plate}, 2—(002) of M_{plate}, 3—(200) of M_{lath}, and 4—(002) of M_{lath}.

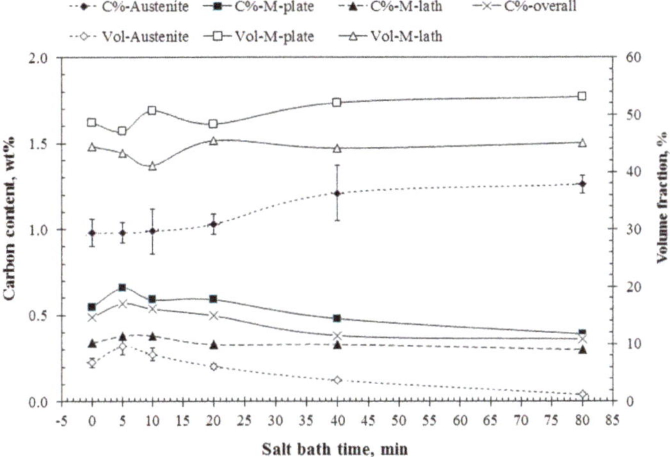

Figure 8. Quantified results of XRD analyses: Effect of austempering time on the volume fraction, carbon content and lattice strain of the matrix phases.

Figure 8 illustrates the effect of austempering time on the variation of volume fraction and carbon concentration of the three matrix constituents. In general, the ferritic M_{plate} and M_{lath} sub-phases account for the majority of volume fraction, whereas the retained austenite is much less. The M_{plate} sub-phase has a higher carbon concentration and a larger lattice strain (see Figure 1) than the M_{lath} sub-phase. In addition, retained austenite shows enriched carbon content, as compared to the ferritic phases.

It is interesting to note the differences between the oil-quenched and short-time austempered samples. Firstly, the 5 min austempering pre-treatment resulted in slightly increased volume fractions of retained austenite from 6.8% to 9.6%, while its carbon concentration remained unchanged at 0.98%. Accordingly, the volume fractions of the M_{plate} and M_{lath} decreased slightly. In addition, the M_{plate} of the austempered sample shows higher carbon concentration (0.57%) than the oil-quenched counterpart (0.49%) along with increased lattice strain from 0.79% to 0.84%. The less carbon concentration in the oil-quenched M_{plate} may be related to simultaneous carbide precipitation as suggested in the SEM observation (e.g., Figure 3c). A similar difference in the carbon concentration is also found in the M_{lath} sub-phase. The 10 min austempered sample also shows superior amount of retained austenite as compared to the oil-quenched sample as well as slightly increased ferrite carbon concentrations, although not as strong as in the 5 min austempered sample.

The austempering time shows pronounced influence on the structural properties. A high carbon content of 0.59–0.66% was retained in the M_{plate} when the sample was austempered for less than 20 min. Then it decreased to 0.48% and 0.39% respectively, when the bath time was increased to 40 and 80 min. For the M_{lath}, its carbon content decreased from 0.38% to 0.30% when the austempering time was increased from 5 to 80 min, which is relatively negligible as compared to the M_{plate}. The more pronounced decarbonisation in the M_{plate} is consistent with the SEM observations that carbide precipitates appeared preferentially in the M_{plate} grains. Meanwhile, the volume fraction of retained austenite decreased from 6.1% to 1.2% when the austempering time was increased from 20 to 80 min, accompanying continuous enrichment in its carbon concentration from 1.03% to 1.26%. The results suggest partial decomposition of the austenite and continuous carbon partitioning between the austenite and ferrite. Finally, the overall carbon content illustrated in Figure 8 stands for the average carbon content in the iron-based matrix, including the martensitic and bainitic ferrites and retained austenite, except for carbide precipitates. The overall carbon content of matrix starts decreasing when the austempering time was increased from 20 to 40 min, which is consistent to the pronounced carbide precipitation.

4. Discussion

4.1. Hardening Mechanism in Austempering Treatment

The results presented reveal that, a short-term isothermal bainitic transformation can be introduced prior to the continuous-cooling martensite transformation, which would favour both the refinement of the overall bainitic/martensitic ferrite microstructure and the maximised hardening. Such austempering has served as a replacement of conventional oil-quenching, promising ultrahigh strength properties, as shown in our recent work [28].

Regarding the hardening mechanisms, our finding that short-time austempering treatment promotes grain refining is consistent with the results of others. For example, previous TEM observations have revealed that, oil-quenched 300 M steel showed relatively lengthy martensitic laths in contrast to the isothermally transformed bainitic laths of the same material [36]. However, our findings differ from the mechanisms suggested by those previous studies, in which the martensite grain refining was attributed to partitioning of austenite grains by pre-formed bainitic ferrite [3,11–15,37]. If such a mechanism were true in the current experiments, the steel would have shown lengthy bainitic ferrite laths or plates to partition the austenite grains. In fact, such length bainitic needles were not observed, as shown in Figures 3b,d, 4a and 5a. Instead, we found that grain coarsening took place when the austempering time exceeded 20 min. The findings suggest that martensite refining took place prior to the growth of the bainitic laths.

Instead of the partitioning mechanism, the current work is more likely explained by the mechanism of carbon spinodal decomposition, in other words, carbon re-distribution in the under-cooled austenite. In this isothermal period, atomic and electronic interactions might happen between the carbon atoms and the metal atoms in the austenite lattice, which results in the heterogeneous distribution of carbon atoms in the austenite lattice. Consequently nanoscale body-centre cubic embryos could nuclear in the carbon-depleting regions of the parent austenite. The spinodal model was proposed by Kang and his group when spinodal decomposition of carbon-rich clusters was evidenced in their TEM investigation [32,33]. In their publications, the formation of carbide-free bainitic ferrite was described as a series of phase transformation sub-processes starting from the localised clustering or spinodal decomposition within under-cooled austenite. The spinodal decomposition required carbon atoms to undergo only short-range motion, which is time-dependent and would be available at the austempering temperature. Then the spinodal decomposition was proposed to trigger the formation of bainite embryos in the resultant carbon-depleted regions, before the embryo growth under the displacive model. The embryo growth proceeds along with simultaneous carbon diffusion across the austenite–ferrite interface to enhance carbon enrichment in the adjacent austenite. Then, subsequent nucleation and growth of carbide precipitates take place in the bainitic ferrite [38,39]. Several other groups have also confirmed the carbon heterogeneity in super-cooled austenite and martensitic/bainitic ferrite [22,26,40–42].

As a result of the spinodal decomposition, the carbon distribution would be more heterogeneous in nano-scale, so that a large number of ferrite nuclei could be generated in the carbon-depletion domains. For the samples being austempered for only 5 min the SEM observations did not reveal any bainitic ferrite, as shown in Figure 3b,d. Instead, the under-cooled austenite transformed to martensite in the subsequent cooling. The resultant refining can be explained by the increased nucleation sites of the martensites as a sequence of the heterogeneous distribution of carbon.

In addition to these, our results suggest that the role of carbon in strengthening can be described in two aspects. In the first, the short-term austempering treatment did not lead to carbide precipitation in the bainitic ferrite. Consequently, the bainitic and martensitic ferrite still remained a super-saturated carbon concentration, leading to the maximum solute hardening. Secondly, the combined XRD and Gaussian multiple peak-fitting analyses have indicated different tetragonal ratios and carbon concentrations of the lath- and plate-shape ferrites, both remaining at high levels of carbon supersaturation, as compared to the oil-quenched martensites. The carbon supersaturation is known to ensure a high level of solute hardening of the martensites.

4.2. Effect of Austempering Time on the Resultant Microstructure

Moreover, the austempering treatment has been found to lead to microstructure evolution in three aspects, namely, re-distribution of carbon, coarsening of bainitic ferrite, and continuous decomposition of retained austenite.

Firstly, the carbon re-distribution includes a continuous decrease of carbon supersaturation in the ferritic matrix, through precipitation and growth of dispersive carbide particles, as shown in Figure 5. More precisely, the carbon re-distribution took place preferentially in the high-carbon plate-shape bainitic ferrite grains, whereas the low-carbon lath-shape ferrite almost remained at the same carbon concentration, as shown in Figure 8. Meanwhile, carbon atoms in the ferrite continued immigration to the adjacent austenite to make the latter even richer in carbon. The increased carbon enrichment has been confirmed in the XRD analysis, as shown in Figure 8. Such heterogeneous distribution of carbon in martensitic/bainitic microstructure has been confirmed by other sophisticated analyses. In a more recent paper, atomic probe tomography of a nanobainitic steel revealed the heterogeneous carbon content in bainitic ferrite, where the ferrite is in close vicinity of the ferrite austenite interface, showed low carbon concentration as compared to the remarkably high carbon concentration in the core ferrite region [43].

Secondly, the longitudinal and transverse dimensions of the bainitic ferrite grains increased with the austempering time. The current experiment results provide evidences on the time-dependent transformation of bainite structure. It is well known that there has long been a controversy on the displacive or shear model and diffusive model of bainite transformation as documented in the literature, e.g., a latest review paper [44]. In the present case, we believe the growth of bainitic ferrite laths was dominated, or at least strongly influenced, by the diffusion of carbon.

In the last, the quantity of retained austenite continued to decrease, and became almost non-detectable when the austempering was 80 min.

Furthermore, we also noticed in the current research that dendritic segregations initiated in steel casting still show significant influence on the chemical homogeneity of the steel even after several rounds of thermal processing. This has been evidenced by the heterogeneous distributions of both the lath and plate martensites in oil-quenched samples and the bainitic and martensitic sub-structures. The segregation would have happened firstly on the substitutional elements, such as Mn, Cr, Ni and Mo. Carbon is known to be attractive to carbide-forming elements and repulsive to non-carbide-forming elements [45,46]. As a result, the heterogeneity of the substitutional elements also influences the distribution of carbon, and consequently may have influenced the transformation kinetics, from the spinodal decomposition of under-cooled austenite to the carbon-diffusion affected ferrite growth. More research attention will be given to this issue later.

5. Conclusions

Short-time austempering treatment of the 56NiCrMoV7 spring steel in a salt-bath time of 5 to 10 min resulted in the maximum hardness values being equivalent to the oil-quenched sample; meanwhile, the resultant microstructure comprised a mixture of fine martensitic and bainitic ferrites and retained austenite. When the austempering time was increased from 20 to 80 min, progressive decrease in the hardness was associated with the evolution of the microstructure, featured by bainitic ferrite coarsening, carbide precipitating inside high-carbon bainitic ferrite and its subsequent decarbonisation.

Carbon partitioning showed significant influence on the hardening in several aspects:

(1) Soaking super-cooled austenite at a temperature above its M_s temperature favours the refining of the transformed ferritic microstructure, which may be related to short-range spinodal decomposition of carbon in the incubation period;

(2) The best hardening state is obtained prior to remarkable carbide precipitation, i.e., when most carbon atoms remain supersaturated in the bct-structured ferrite;

(3) Following longer austempering time, the bainitic ferrite becomes increasingly decarbonised through continuous carbon clustering and carbide precipitation.

Acknowledgments: The authors acknowledge that the research was partially sponsored by Innovate UK (formerly Technology Strategy Board) of the UK government through Smart Award No. 720113. Tinsley Bridge Limited is acknowledged for providing the sample steel and colaboration in the Smart Award project.

Author Contributions: Matthew Kitchen participated in the heat treatments, sample preparation, and SEM analyses; Shahriar Abubakri participated in sample preparation, OPM, hardness testing, and XRD experiments; Quanshun Luo led the research, participated in the heat treatments, X-ray diffraction quantification and SEM, and wrote the paper.

Conflicts of Interest: The authors declare no conflict of interest.

References

1. Li, Z.; He, Z.Q.; Jin, J.J.; Zhong, P. *Development of Aeronautical Ultra-High Strength Steels*; National Defence Industry Press: Beijing, China, 2012.

2. Krauss, G. Deformation and fracture in martensitic carbon steels tempered at low temperatures. *Metall. Mater. Trans. B* **2001**, *32*, 205–221. [CrossRef]

3. Tomita, Y. Development of fracture toughness of ultrahigh strength, medium carbon, low alloy steels for aerospace applications. *Int. Mater. Rev.* **2000**, *45*, 27–37. [CrossRef]

4. Caballero, F.G.; Bhadeshia, H.K.D.H.; Mawell, K.J.A.; Jones, G.D.; Brown, P. Design of novel high strength bainitic steels. *Mater. Sci. Technol.* **2001**, *17*, 512–522. [CrossRef]

5. Caballero, F.G.; Bhadeshia, H.K.D.H.; Mawell, K.J.A.; Jones, D.G.; Brown, P. Very strong low temperature bainite. *Mater. Sci. Technol.* **2002**, *18*, 279–284. [CrossRef]

6. Caballero, F.G.; Bhadeshia, H.K.D.H. Very strong bainite. *Curr. Opin. Solid State Mater. Sci.* **2004**, *8*, 251–257. [CrossRef]

7. Garcia-Mateo, C.; Caballero, F.G. Ultra high strength bainitic steels. *ISIJ Int.* **2005**, *45*, 1736–1740. [CrossRef]

8. Kang, M.K.; Zhu, M.; Zhang, M.X. Mechanism of bainite nucleation in steel, iron and copper alloys. *J. Mater. Sci. Technol.* **2005**, *21*, 437–444.

9. Kang, M.K.; Zhu, M. Stabilization of austenite in quenched alloy steels. *Acta Metall. Sin.* **2005**, *41*, 673–679.

10. Wang, T.S.; Li, X.Y.; Zhang, F.C.; Zheng, Y.Z. Microstructures and mechanical properties of 60Si2CrVA steel by isothermal transformation at low temperature. *Mater. Sci. Eng.* **2006**, *438–440*, 1124–1127. [CrossRef]

11. Malakondaiah, G.; Srinivas, M.; Rao, P.R. Ultrahigh-strength low-alloy steels with enhanced fracture toughness. *Prog. Mater. Sci.* **1997**, *42*, 209–242. [CrossRef]

12. Rao, T.V.L.N.; Dikshit, S.N.; Malakondaiah, G.; Rap, P.R. On mixed upper bainite-martensite in an AISI 4330 steel exhibiting an uncommonly improved strength-toughness combination. *Scr. Metall. Mater.* **1990**, *24*, 1323–1328. [CrossRef]

13. Park, K.T.; Kwon, H.J. Interpretation of the strengthening of steel with lower bainite and martensite mixed microstructure. *Met. Mater. Int.* **2001**, *7*, 95–99. [CrossRef]

14. Abbaszadeh, K.; Kheirandish, S.; Saghafian, H. The effect of lower bainite volume fraction on tensile and impact properties of D6AC medium carbon low alloy ultrahigh strength steel. *Iran. J. Mater. Sci. Eng.* **2010**, *7*, 31–38.

15. Sharma, S.; Sangal, S.; Mondal, K. Development of new high-strength carbide-free bainitic steels. *Metall. Mater. Trans. A* **2011**, *42*, 3921–3923. [CrossRef]

16. Safi, S.M.; Givi, M.K.B. A new step heat treatment for steel AISI 4340. *Met. Sci. Heat Treat.* **2014**, *56*, 79–81. [CrossRef]

17. Lan, H.F.; Du, L.X.; Zhou, N.; Liu, X.H. Effect of austempering route on microstructural characterization of nanobainitic steel. *Acta Metall. Sin.* **2014**, *27*, 19–26. [CrossRef]

18. Young, C.H.; Bhadeshia, H.K.D.H. Strength of mixtures of bainite and martensite. *Mater. Sci. Technol.* **1994**, *10*, 209–214. [CrossRef]
19. Kang, M.K.; Ai, Y.L.; Zhang, M.X.; Yang, Y.Q.; Zhu, M.; Chen, Y. Carbon content of bainite ferrite in 40CrMnSiMoV steel. *Mater. Chem. Phys.* **2009**, *118*, 438–441. [CrossRef]
20. Garcia-Mateo, C.; Jimenez, J.A.; Yen, H.W.; Miller, M.K.; Morales-Rivas, L.; Kuntz, M.; Ringer, S.P.; Yang, J.R.; Caballero, F.G. Low temperature bainitic ferrite: Evidence of carbon super-saturation and tetragonality. *Acta Mater.* **2015**, *91*, 162–173. [CrossRef]
21. Garcia-Mateo, C.; Caballero, F.G.; Miller, M.K.; Jimenez, J.A. On measurement of carbon content in retained austenite in a nanostructured bainitic steel. *J. Mater. Sci.* **2012**, *57*, 1004–1010. [CrossRef]
22. Speer, J.G.; Matlock, D.K.; De Cooman, B.C.; Schroth, J.G. Carbon partitioning into austenite after martensite transformation. *Acta Mater.* **2003**, *51*, 2611–2622. [CrossRef]
23. Edmonds, D.V.; He, K.; Rizzo, F.C.; De Cooman, B.C.; Matlock, D.K.; Speer, J.G. Quenching and partitioning martensite—A novel steel heat treatment. *Mater. Sci. Eng. A* **2006**, *438–440*, 25–34. [CrossRef]
24. Rong, Y. Advanced Q–P–T steels with ultrahigh strength-high ductility. *Acta Metall. Sin.* **2011**, *47*, 1483–1489.
25. Li, H.Y.; Lu, X.W.; Li, W.J.; Jin, X.J. Microstructure and mechanical properties of an ultrahigh-strength 40SiMnNiCr steel during the on-step quenching and partitioning process. *Metall. Mater. Trans. A* **2010**, *41*, 1284–1300. [CrossRef]
26. Hsu, T.Y. Carbon diffusion and kinetics during the lath martensite formation. *J. Phys. IV Fr.* **1995**, *5*, C8-351–C8-354.
27. Luo, Q.; Kitchen, M.; Patel, V.; Magowan, S. Carbon partitioning and structure evolution in the hardening treatments of high strength steel. In Proceedings of the 20th Congress of International Federation for Heat Treatment and Surface Engineering, Beijing, China, 23–25 October 2012; pp. 111–117.
28. Luo, Q.; Kitchen, M.; Patel, V.; Filleul, M.; Owens, D. Partial isothermally treated low alloy ultrahigh strength steel with martensitic/bainitic microstructure. In *HSLA Steels 2015, Microalloying 2015 & Offshore Engineering Steels 2015*; John Wiley & Sons: Hoboken, NJ, USA, 2015; pp. 433–438.
29. Garg, A.; McNelley, T.R. Estimation of martensite carbon content in as-quenched AISI 52100 steel by X-ray diffraction. *Mater. Lett.* **1986**, *4*, 214–218. [CrossRef]
30. Abbaschian, R.; Abbeschian, L. *Physical Metallurgy Principles, Reed-Hill RE*, 4th ed.; Cengage Learning: Stanford, CA, USA, 1994.
31. Luo, Q. A new XRD method to quantify plate and lath martensites of hardened medium-carbon steel. *J. Mater. Eng. Perform.* **2016**, *25*, 2170–2179. [CrossRef]
32. Kang, M.K.; Yang, Y.Q.; Wei, Q.M.; Yang, Q.M.; Meng, X.K. On the prebainitic phenomenon in some alloys. *Metall. Mater. Trans. A* **1994**, *25*, 1941–1946. [CrossRef]
33. Wu, X.L.; Zhang, X.Y.; Meng, X.K.; Kang, M.K.; Yang, Y.Q. Formation of carbon-poor regions during pre-bainitic transformation. *Mater. Lett.* **1995**, *22*, 141–144. [CrossRef]
34. Khan, S.A.; Bhadeshia, H.K.D.H. The bainite transformation in chemically heterogeneous 300M high-strength steel. *Metall. Trans. A* **1990**, *21*, 859–875. [CrossRef]
35. Zhang, X.Y.; Kang, M.K.; Wu, X.L.; Chen, D.M.; Han, D. Study on several carbide variants in the low bainite of 65Si2MnWA steel by TEM. *Chin. Sci. Bull.* **1994**, *39*, 1583–1584.
36. Luo, C.P.; Liu, J. Crystallography of lath martensite and lower bainite in alloy steels. *Mater. Sci. Eng. A* **2006**, *438–440*, 149–152. [CrossRef]
37. Yang, F.B.; Bai, B.Z.; Liu, D.Y.; Chang, K.D.; Wei, D.Y.; Fang, H.S. Microstructure and properties of a carbide-free bainite/martensite ultrahigh strength steel. *Acta Metall. Sin.* **2004**, *40*, 296–300.
38. Kang, M.K.; Yang, Y.Q.; Zhang, X.Y.; Sun, J.L.; Jia, F.S.; Wu, X.L. Bainitic transformation in silicon-containing steels. *Acta Metall. Sin.* **1996**, *32*, 897–903.
39. Kang, M.K.; Zhang, M.X.; Zhu, M. In-situ observation of bainite growth during isothermal holding. *Acta Mater.* **2006**, *54*, 2121–2129. [CrossRef]
40. Hsu, T.Y.; Li, X.M. Diffusion of carbon during the formation of low-carbon martensite. *Scr. Metall.* **1983**, *17*, 1285–1288. [CrossRef]
41. Lawrynowicz, Z. Carbon partitioning during bainite transformation in low alloy steels. *Mater. Sci. Technol.* **2002**, *18*, 1322–1324. [CrossRef]
42. Caballero, F.G.; Miller, M.K.; Clarke, A.J.; Garcia-Mateo, C. Examination of carbon partitioning into austenite during tempering of bainite. *Scr. Mater.* **2010**, *63*, 442–445. [CrossRef]

43. Timokhina, I.B.; Beladi, H.; Xiong, X.Y.; Adachi, Y.; Hodgson, P.D. Nanoscale microstructure characterization of a nanobainitic steel. *Acta Mater.* **2011**, *59*, 5511–5522. [CrossRef]
44. Fielding, L.C.D. The bainite controversy. *Mater. Sci. Technol.* **2013**, *29*, 383–399. [CrossRef]
45. Gavriljuk, V.G.; Shanina, B.D.; Berns, H. On the correlation between electron structure and short range atomic order in iron-based alloys. *Acta Mater.* **2000**, *48*, 3879–3893. [CrossRef]
46. Shanint, B.D.; Gavriljuk, V.G.; Konchits, A.A.; Kolesnik, S.P. The influence of substitutional atoms upon the electron structure of the iron-based transition metal alloys. *J. Phys.* **1998**, *10*, 1825–1838.

Article

Effects of Q&P Processing Conditions on Austenite Carbon Enrichment Studied by In Situ High-Energy X-ray Diffraction Experiments

Sébastien Yves Pierre Allain [1,*], Guillaume Geandier [1], Jean-Christophe Hell [2], Michel Soler [2], Frédéric Danoix [3] and Mohamed Gouné [4]

[1]   Institut Jean Lamour, UMR CNRS-UL 7198, 54011 Nancy, France; guillaume.geandier@univ-lorraine.fr
[2]   Maizières Automotive Products, Arcelormittal Maizières Research SA, 57283 Maizières les Metz, France; jean-christophe.hell@arcelormittal.com (J.-C.H.); michel.soler@arcelormittal.com (M.S.)
[3]   Groupe de Physique des Matériaux, UMR 6634, Normandie University, UNIVROUEN, INSA Rouen, CNRS, 76801 Rouen, France; frederic.danoix@univ-rouen.fr
[4]   Institut de Chimie de la Matière Condensée de Bordeaux, UPR 9048, 33608 Pessac, France; mm.goune@gmail.com
*   Correspondence: sebastien.allain@univ-lorraine.fr; Tel.: +33-383-584-377

Received: 13 May 2017; Accepted: 16 June 2017; Published: 22 June 2017

Abstract: We report the first ultra-fast time-resolved quantitative information on the quenching and partitioning process of conventional high-strength steel by an in situ high-energy X-ray diffraction (HEXRD) experiment. The time and temperature evolutions of phase fractions, their carbon content, and internal stresses were determined and discussed for different process parameters. It is shown that the austenite-to-martensite transformation below the martensite start temperature *Ms* is followed by a stage of fast carbon enrichment in austenite during isothermal holding at both 400 and 450 °C. The analysis proposed supports the concurrent bainite transformation and carbon diffusion from martensite to austenite as the main mechanisms of this enrichment. Furthermore, we give evidence that high hydrostatic tensile stresses in austenite are produced during the final quenching, and must be taken into account for the estimation of the carbon content in austenite. Finally, a large amount of carbon is shown to be trapped in the microstructure.

Keywords: steel; martensite; bainite; Q&P; synchrotron; HEXRD; TRIP

---

1. Introduction

Quenching & Partitioning (Q&P) is a new annealing route proposed to produce third generation advanced high-strength steels (AHSS) for the automotive sector. The Q&P annealing cycle consists, after an austenitic soaking, of an interrupted quenching at a temperature—the quenching temperature (QT), lower than the martensite start temperature *Ms*, but higher than the martensite finish temperature *Mf*—to reach a partial martensitic transformation. This quench is followed by an isothermal holding at a temperature called the partitioning temperature (PT). During this second step, it was unambiguously shown that carbon can diffuse from martensite to untransformed austenite, from measurements at the atomic scale [1,2]. In a seminal article in 1960 published in Nature, Matas S. and Hehemann M. F. had already shown that the tempering of martensite leads to a carbon-enriched austenite [3]. In common steels, this process is largely inhibited by carbide precipitation (the so-called tempering process). A judicious choice of alloying elements (Si, Al) and partitioning temperatures permits reduction of transition carbide, and even cementite, precipitation, depending on the holding temperature [4–6]. As a result, austenite can be significantly enriched in carbon and thus stabilized at room temperature (RT). This is the so-called partitioning process. The cycle ends with a cooling sequence, during which a certain amount of fresh martensite can form again. Accordingly, Q&P steels are thus nanostructured austenite/martensite duplex steels with a recovered/tempered martensitic matrix. In most cases, they can show an efficient transformation induced plasticity (TRIP) effect upon straining. A typical Q&P annealing cycle is mainly characterized by its QT and its PT, as depicted in Figure 1. It should be mentioned that the PT is not necessarily higher than the QT, and that carbon partitioning can take place along various time–temperature schemes, including isothermal holding at the QT and even during continuous cooling [7].

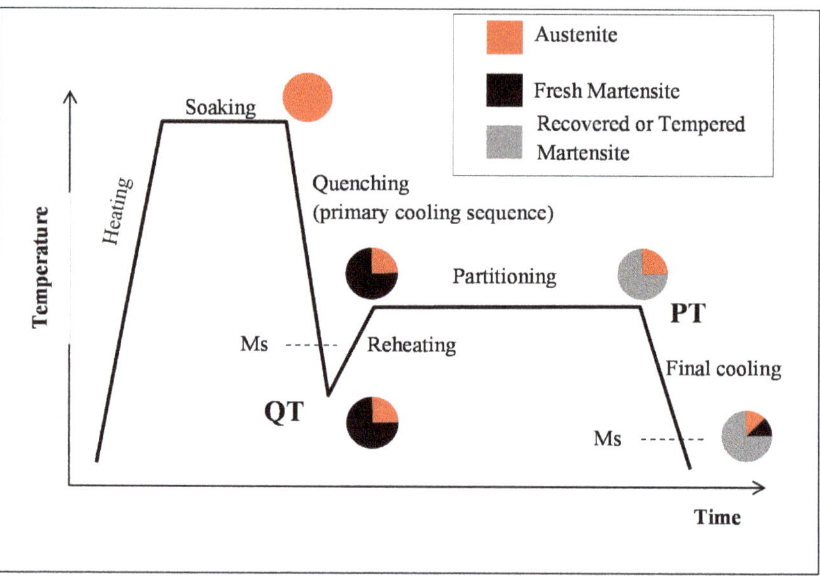

Figure 1. Typical time–temperature scheme to produce Quenching & Partitioning (Q&P) Transformation Induced Plasticity (TRIP) steels after a full austenitic soaking, and corresponding microstructures, according to [4–6]. The quench temperature (QT) and the partitioning temperature (PT) characterizing the cycle are positioned on the cycle.

As in more conventional TRIP steels with a ferritic matrix, the key to benefiting from an efficient TRIP effect relies on the thermo-mechanical stability of retained austenite islands [8]. Their stabilities primarily depend on their local carbon content [9,10], but also depend on their size, environment, and residual internal stress [11]. The question of austenite carbon enrichment remains of prime interest for the final properties of the steel. Recent studies highlight that it cannot be simply deduced from the sole partitioning process between austenite and martensite, but that at least two additional mechanisms must be considered, as they both potentially affect the carbon distribution between phases in Q&P steels—namely carbide precipitation and bainite formation.

The first mechanism is carbide precipitation, as it is now well admitted that precipitation of carbides interacts strongly with partitioning processes [6]. For HajyAkbary et al. [12], ε carbides are necessary for complete carbon partitioning, as they prevent possible bainitic transformations. On similar alloys, Pierce et al. [13] on the contrary have recently reported precipitation of η carbides during the partitioning process, which has been identified by transmission electron microscopy (TEM) and Mossbauer spectroscopy (MS). For Toji et al. [14], carbides appearing during partitioning are θ carbides (cementite), based on atom probe tomography (APT) results. Their observations are sustained with an original reassessed thermo-kinetic model. Carbon clustering in martensite has also been reported by Thomas et al. [15] in a highly-alloyed system. Even if the nature of the carbide remains the bone of contention, all these recent studies lead to the conclusion that a certain fraction of carbon must also be trapped in recovered martensitic laths, preventing a complete carbon partitioning between martensite and austenite, thus limiting austenite carbon enrichment.

Secondly, the partitioning step of the Q&P process is often conducted at temperatures which enable bainite formation in retained austenite [6,16–18]. The formation of bainite could explain, to some extent, carbon enrichment in austenite, as in carbide-free bainitic steels [19–21]. Using in situ HEXRD (high-energy X-ray diffraction) experiments, Nishiwaka et al. observed an evolution of the fraction of bainite during the partitioning step in hypereutectoid steels [22]. On more conventional TRIP steels, the same group showed how the bainitic transformation could be hindered by a prior deformation of austenite at a high temperature [10]. HajyAkbary et al. also reported on volume changes using a dilatometer during partitioning, which they attributed to bainitic transformations [12]. They claimed that this latter transformation was in fact unavoidable in the studied Q&P steels, and correlated it with possible ε carbide precipitation. In a recent paper, the present authors investigated the Q&P process in low-carbon TRIP steels by in situ HEXRD experiments coupled with a diffusion model [23]. Carbon enrichment in austenite was shown to be the result of the competition between the direct partitioning process and the carbide-free bainitic transformation. They also showed that a significant fraction of carbon remains trapped in martensite, supporting the conclusions of Thomas et al. [15].

Q&P microstructures are thus intricate, resulting from the conjunction of different mechanisms (two martensitic transformations, carbon precipitation and segregation in martensite, partitioning between martensite and austenite, possible bainitic transformation, etc.), which depend on the processing conditions. The respective kinetics of these elementary mechanisms affect the final microstructures and thus are of prime interest for final properties [8,10,16]. In this paper, the effect of the Q&P processing conditions were investigated by means of in situ HEXRD experiments, as in the prior works of [24] or [10], and further analyzed on the basis of the unbiased carbon mass balance allowed by this method [23].

2. Materials and Methods

2.1. Studied Alloy

The studied steel was Fe–0.295 C–2.52 Mn–1.43 Si–0.81 Cr (wt %, as for all compositions given in this paper). The same alloy has been used in previous studies on carbide-free bainitic transformations by Hell et al. [19,20], and on Q&P [23]. All the details about the sample preparation can be found in these prior papers. The *Ms* value of the studied steel, 295 °C, was measured during the in situ investigations, and was consistent with the value calculated with the model developed by Van Bohemen (298 °C) [25,26].

2.2. Diffraction Set-Up and Data Processing

The HEXRD experiments were carried out on the European Synchrotron Radiation Facility (ESRF) ID15B line (Grenoble, France) under powder diffraction configuration. The high-energy monochromatic beam (E = 87 keV, λ = 0.14 nm) allowed working in transmission mode, and the association with a fast 2D Perkin-Elmer detector (PerkinElmer, Waltham, MA, USA) enabled high acquisition rates (10 Hz) needed to study "real-time" processes on bulk samples. The detector was positioned about 1 m behind the sample, giving access to full Debye-Scherrer rings with a maximum 2θ angle of 12°.

The 2D diffraction patterns produced during the experiments were integrated circularly using Fit2D software [27]. The deduced 1D diffractograms (intensity vs. 2θ) were analyzed with a full Rietveld refinement procedure. Diffraction peaks were modeled by pseudo-Voigt functions using FullProf software [28] with 16 degrees of freedom for each record (background, phase fraction, lattice parameters, shape of peaks, and temperature effects). An example of such post-treatment is detailed in [23].

For all the different experimental spectra, two phases could be unambiguously identified on diffraction patterns: a face-centered cubic (fcc) phase, corresponding to austenite; and a body-centered (bc) phase. The latter could correspond to either body-centered tetragonal (bct) martensite or body-centered cubic (bcc) bainitic ferrite. As it is was not yet possible to isolate the possible contribution of each based on tetragonal distortion, they were both considered as body-centered cubic. As a consequence, during Rietveld refinement procedures, the lattices of the two phases were considered as cubic (Fm3m for austenite, and Im3m for martensite/bainite). Therefore, only mean volume effects of the composition or hydrostatic stress evolutions in phases were analyzed so far. The carbon mass balances were established considering the sole austenite lattice parameter evolutions.

The careful study of main diffraction peak shoulders revealed the possible presence of a third phase after the first martensitic transformation. The corresponding minor peaks were similar to those reported recently by [29] in nanobainitic steels, and could correspond to η carbides [30] (characteristic single and isolated diffraction peak located at about 5.08°). The precipitation of such carbides in Q&P steels has already been suggested by [13] on the basis of MS and TEM experiments. Other kinds of carbides (cementite or epsilon carbide in particular) were not able to explain the experimental diffraction patterns. However, because of the signal-to-noise ratio, it has not been possible to conclude definitively on the nature of the phase, nor to quantify it. These carbides were thus neglected in the following data processing procedure.

All additional details about the diffraction set-up and data processing can be found in [23], as the experiments were realized under the same conditions.

2.3. Quenching & Partitioning (Q&P) Processing Conditions

The Q&P cycles were conducted using a commercial Instron electro-thermal mechanical testing (ETMT) (Instron, Norwood, MA, USA) thermomechanical device. The samples were heated up by the Joule effect and naturally cooled down through cold jaws. This explained, to some extent, transient regimes observed when changing heating or cooling rates. Temperature was measured and regulated with a thermocouple welded on the sample as close as possible to the analyzed region. This particular set-up permitted measuring microstructure evolutions all along the cycle, and capturing fast phenomena, especially close to the primary cooling sequence and during reheating (about 1500 2D X-ray diffraction spectra were acquired for each Q&P cycle).

Three different Q&P cycles, with a different QT and PT, as discussed below, were investigated. The samples were first heated at 900 °C for 200 s with a heating rate of 10 °C/s to reach a fully austenitic state. They were then cooled rapidly at an initial cooling rate of 50 °C/s in order to avoid pro-eutectoid ferritic and bainitic transformations before primary martensitic transformation, as shown by Hell et al. [19,20]. Due to the cooling method, a slight change in the cooling rate was observed at the *Ms* temperature because of the latent transformation heat.

Two different QTs were chosen to optimize the fraction of retained austenite, according to the original approach of Speer et al. [4–6]. For the investigated alloy, the predicted retained austenite fraction as a function of the QT after the long partitioning time is represented in Figure 2. The calculation relied on a Carbon Constraint Equilibrium (CCE) assumption and the empirical Van Bohemen et al. [25,26] equations for martensitic transformations (initial and final). The CCE is an interface condition between martensite and austenite assuming that only the chemical potential of carbon must be equilibrated in both phases, and that the interface is fixed as a consequence [4–6]. The validity of such mean-field calculations is often discussed in the literature [16] and is considered as an upper-bound by some authors [12]. The treatment QT = 230 °C/PT = 400 °C was considered as the reference in the following, as it corresponded to the higher retained austenite amount (22%). Note that this treatment has already been studied and discussed by Allain et al. [23]. In the present paper, two additional Q&P cycles were investigated for comparison. The first (second cycle) had a lower QT (QT = 200 °C/PT = 400 °C), for which the Speer et al. approach predicted a lower fraction of retained austenite (12%) with a higher carbon content, and thus higher stability. The second (third cycle) had a higher PT (QT = 230 °C/PT = 450 °C), where the predicted retained austenite fraction at RT was similar to the reference cycle after long partitioning [16]. As differences in bainitic transformations were expected [12], the final experimental austenite fractions may have differred, highlighting the influence of the bainitic transformation, if present, on the final microstructure at RT. For all considered cycles, the duration of the partitioning step was about 250 s, and the final cooling achieved with a mean cooling rate of 20 °C/s.

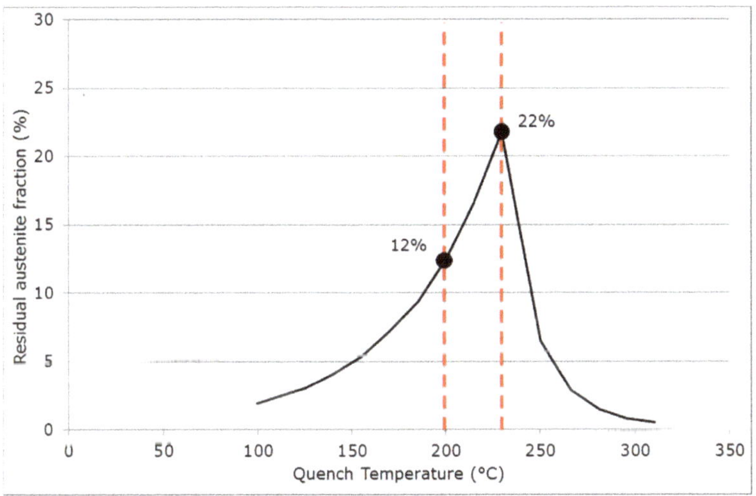

Figure 2. Retained austenite at room temperature (RT) after Q&P as a function of QT in the studied steel according to the classical approach of Speer at al. (Carbon Constraint Equilibrium (CCE) assumption and model of Van Bohemen et al. for martensitic transformation kinetics) [4–6,25,26]. The dashed lines represent both QTs addressed in the present study.

3. Results

3.1. Evolution of Phase Fractions

Figure 3a shows the evolution of the temperature and measured phase fractions as a function of time for the reference cycle: QT = 230 °C/PT = 400 °C. The figure is restricted to the most interesting part of the cycle, i.e., austenite decomposition after soaking. The evolution can be decomposed into six steps, which have been described in [23], and are separated in Figure 3 by red dashed lines (for clarity, the first 50 s have been detailed in Figure 3b):

(i)   During the first cooling step, the alloy remained fully austenitic above *Ms*.
(ii) Primary martensitic transformation below initial *Ms*: The initial *Ms* temperature was estimated to be 295 °C from an extrapolation of the transformation kinetics. The final fraction of martensite before reheating was 65% (note that at this stage, the bcc phase was implicitly but obviously identified as martensite).
(iii) Reheating step, with a duration of a few seconds, because of the regulation procedures of the heating device: The duration of this transient regime was lower than 5 s, before temperature stabilization for the partitioning step. Figure 3b reveals that during the reheating step, the phase fractions remained constant. For all studied conditions, the phase fractions remained constant during reheating up to about 370 °C.
(iv) Partitioning step: Above 370 °C, a significant increase in the bcc phase fraction was observed during the partitioning step. The kinetics were initially fast during the first 50 s, progressively

slowing down. Due to the experimental limitations in revealing tetragonality, the exact nature of this bcc phase was debatable as it could have corresponded to either bainite and/or athermal martensite. Indeed, there was strong evidence of both bainite and athermal martensite formation below *Ms* [7,31,32]. It was even stated that this isothermal product was neither purely martensitic nor purely bainitic [33]. One could even consider the simultaneous formation of both athermal martensite and bainite as an alternative. Indeed, the formation of athermal martensite could have a strong accelerating effect on the subsequent bainite formation by providing a higher density of potential nucleation sites. As a consequence, below *Ms*, the nature of the transformation products during isothermal holding was unclear. In addition, lower bainite and athermal martensite exhibited morphological similarities. The morphological criteria used by Somani et al. to distinguish bainite from athermal martensite (laths with wavy boundaries and ledge-like protrusions) [7] was discussed very recently by [34] in a convincing way. According to their analysis, below *Ms*, the driving force for bainite nucleation is so high that small units of bainite may grow from the initial martensite laths in the form of ledge-liked protrusions. Furthermore, bainitic ferrite can grow from the prior athermal martensite, maintaining a similar orientation relationship that could have contributed to the formation of the ledge-like protrusions that could have, in turn, given rise to a wavy appearance of the boundaries.

In the present study, it is worth noting that the austenite-to-martensite transformation below *Ms* was followed by a stagnant stage during re-heating, during which no phase transformation occurred until 370 °C. The growth of the bcc phase started at around 370 °C, a temperature which was much higher than the *Ms* temperature (295 °C). It was thus difficult to imagine that athermal martensite could be formed above 370 °C, and not between *Ms* and 370 °C, where the driving force for athermal transformation was much higher. Furthermore, it was shown that ferritic bainite formation is favoured at high partitioning temperatures above 400 °C [7,34]. That is why the bcc phase formed during the partitioning step was considered as bainite.

(v)   During the first part of the final cooling step, the fraction of the bcc phase identified as bainite remained constant down to 120 °C.

(vi) Final martensitic transformation, evidenced by a 2% increase in the bcc phase fraction below 120 °C: Note that again, similarly to step (ii), the increase in the bcc phase was obviously attributed to martensite formation. The final fraction of retained austenite at RT was about 27%.

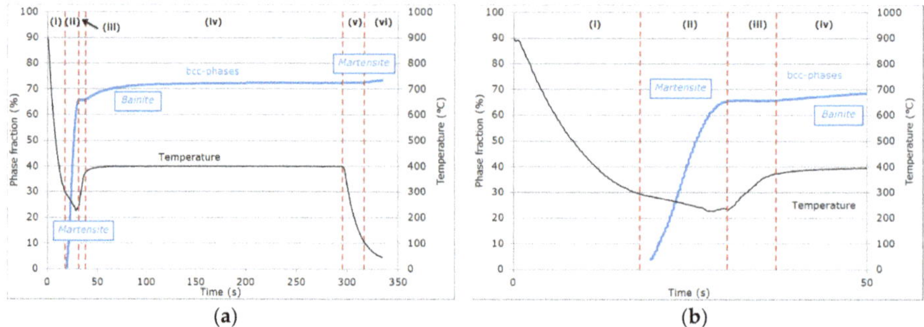

(a)   (b)

Figure 3. (**a**) Evolution of phase fractions and temperature as a function of time measured along the reference cycle: QT = 230 °C/PT = 400 °C. (**b**) Focus is on the first 50 s, covering the first four steps detailed in the text. As discussed in [23], the maximum error made on fraction measurements was about ±1%.

Similar evolutions were observed at a higher PT or at a lower QT. Figure 4 shows the evolution of bcc phase fractions and temperature as a function of time measured along the two other Q&P cycles (defined by: QT = 230 °C/PT = 450 °C and QT = 200 °C/PT = 400 °C).

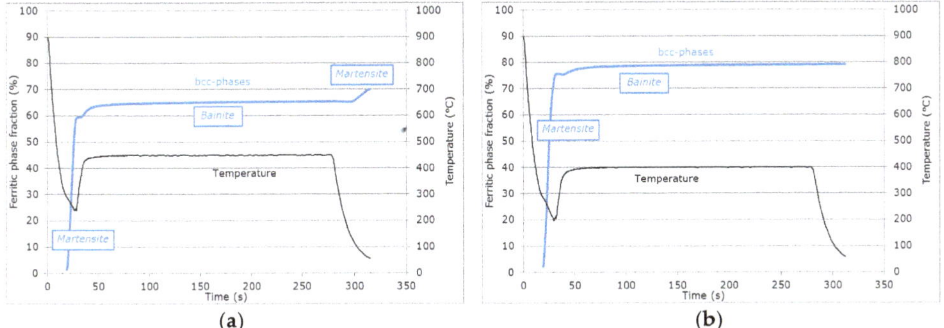

Figure 4. Evolution of phase fractions and temperatures as a function of time measured along two different Q&P cycles defined by: (**a**) QT = 230 °C/PT = 450 °C, and (**b**) QT = 200 °C/PT = 400 °C.

Compared to the reference cycle, the bainite fraction formed was similar (+7%) at a higher PT (PT = 450 °C). Nevertheless, the initial martensitic fraction was slightly lower (58%), as the QT reached was slightly higher (237 °C). Due to the fast transformation kinetics, a minor deviation in the QT can have a significant effect on phase fractions. As the secondary martensitic transformation started at 150 °C, the final martensite fraction (+5%) was higher than in the reference case. Finally, the final fraction of retained austenite at RT was about 30%.

At a lower QT (200 °C), the bainite fraction formed during partitioning was lower (+2%). No final martensitic transformation was observed during final cooling (steps (v) and (vi)). The absence of a final martensitic transformation was expected from the model of Speer et al. The final fraction of retained austenite at RT was about 21% for the cycle.

3.2. Evolution of Austenite Lattice Parameter

The evolution of the austenite lattice parameter is complex along a Q&P cycle, as the lattice thermal expansion varies all along the thermal cycle. This thermal expansion was not constant as in studies carried out on isothermal carbide-free bainitic transformations, for instance. Other possible mechanisms affecting the lattice parameter evolution, such as changes in stress state and in chemical compositions, were thus partially screened. To overcome this intricacy, the expected contribution of thermal expansion needed to be subtracted from the raw measurements (as previously done in [23,24,35]). This permitted isolating the mechanical and chemical effects on the lattice parameters. To do so, the mean thermal expansion coefficient was measured during the primary cooling between 800 and 320 °C (step (i)). In this temperature range, the alloy was fully austenitic and could be considered as stress-free and homogeneous with the nominal chemical composition. The experimental value found for the reference cycle ($2.526 \times 10^{-5} \cdot \mathrm{K}^{-1}$) was close to the value reported by Lu et al. ($2.326 \times 10^{-5} \cdot \mathrm{K}^{-1}$) [36] for pure austenitic iron.

The austenite lattice parameter minus the thermal expansion contribution along the cycle will be referred to as Δa_γ in the following. For the reference cycle, the evolution of Δa_γ is represented in Figure 5a as a function of time. By doing so, it was assumed implicitly in this work that the thermal expansion coefficient determined above could be extrapolated at lower temperatures (i.e., considered as independent of any possible chemical composition change, despite the work of [37]) and as independent also of temperature despite the work of [38]). For the studied carbon enrichments and partitioning temperature ranges, these non-linear effects would not affect carbon balances between phases. Nevertheless, they would lead to overestimating the hydrostatic stresses measured in austenite at RT, as discussed below.

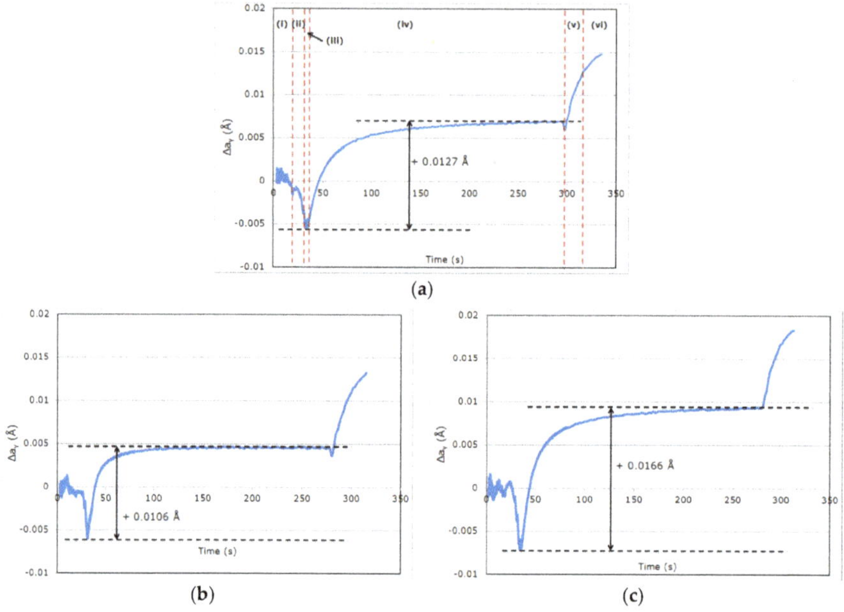

Figure 5. Evolution of Δa_γ as a function of time: (**a**) Q&P reference cycle, (**b**) QT = 230 °C/PT = 450 °C cycle, and (**c**) QT = 200 °C/PT = 400 °C cycle. The different steps discussed in the text are indicated in Figure 5a. In each figure, the increase in the lattice parameter during step (iv) is reported. This increase was attributed to a carbon enrichment in austenite due to both partitioning and bainitic transformations. The error made on the lattice parameter measurements was about ±0.0002 Å after Fullprof analysis.

By construction, Δa_γ was thus constant in step (i), i.e., during initial cooling down to Ms temperature (295 °C at t ~18 s). Below Ms, in step (ii), a drop in Δa_γ was observed during the martensitic transformation. Bruneseaux et al. [39] had already reported such a drop, and related it to second-order internal compressive stresses in austenite, induced by the martensitic transformation (above about 30% martensite). The total decrease in the lattice parameter due to the compressive stress state was −0.0055 Å. As the austenite lattice parameter was 3.6163 Å at t ~35 s at the end of step (ii), the relative lattice volume change was about −0.50%. At 230 °C, this corresponded to a hydrostatic pressure of about −750 MPa, assuming a bulk modulus K = 150 GPa (after the extrapolations of Ghosh and Olson) [40].

During reheating (step (iii)), Δa_γ remained constant again, indicating that the austenite lattice parameter increased only because of thermal expansion.

During partitioning (step (iv)), Δa_γ increased rapidly. This revealed mechanical and/or composition evolutions in austenite. The increase was fast during the first 50 s of partitioning and then sluggish during the rest of the step. The total increase in the lattice parameter was about +0.0127 Å after 250 s, as indicated in Figure 5a. As discussed in our previous work [23], this increase was likely to be due only to carbon enrichment. The relaxation of internal stresses during reheating was ruled out from the analysis as so far, the expected kinetics of such a process were far slower than the observed evolution [41]. Moreover, the full width at half maximums (FWHMs) of austenite peaks, regardless of their Bragg's angles, did not evolve during the partitioning step, which confirmed a low plastic activity during this stage, contrary to what happened in steps (ii) and (iv).

Considering Equation (1) from [23] derived from the work of Toji et al. [18], the relative increase in the carbon content in austenite was estimated at 0.40%. As the initial austenite composition was 0.30% (nominal composition of the alloy), the final carbon content was thus 0.70% at the end of the partitioning step. The new *Ms* temperature of the austenite was then 170 °C, which was consistent with the value observed in Figure 3a (about 120 °C). As explained in [23], the increase in the austenite lattice parameter was strongly time-correlated to the evolution of the bainitic phase fraction. Nevertheless, the transformation was not sufficient alone to explain the measured carbon enrichment in austenite. Both the partitioning and bainitic transformations were thus responsible for this increase.

During final cooling (steps (v) and (vi)), the austenite lattice parameters Δa_γ kept increasing with temperature. To the authors' knowledge, such an increase has never been reported so far in the literature. Accounting for the fast final cooling rate, it is unlikely to be explained by a diffusion controlled mechanism, i.e., by an evolution of the austenite composition. On the contrary, as the microstructure was heterogeneous, the difference in thermal expansion coefficients between the bcc phases and austenite (eigenstrain) could induce second-order internal stresses. This mechanism is well known in composites with a metal or polymer matrix [42,43]. At RT, the total increase in the austenite lattice parameter during final cooling was +0.0078 Å, which corresponded to an additional hydrostatic stress state of about 1175 MPa, assuming a bulk modulus of 177 GPa. As this calculation was based on a linear coefficient of thermal expansion over the whole range of temperatures, this estimation was probably overestimated. At RT, Δa_γ was 0.0148 Å. As 0.0127 Å could be explained by the sole carbon enrichment in austenite, the total residual stress in austenite was only about 295 MPa. This corresponded to the sum of the internal stresses generated during the initial martensitic transformation (compression) and final cooling (tension)

The austenite lattice parameter at RT was thus affected by three different mechanisms in Q&P steels. First, the initial martensitic transformation during step (ii) induced compressive hydrostatic stresses in austenite. Second, carbon enrichment during step (iv) caused a dilatation of the lattice. At last, during final cooling (steps (v) and (vi)), differences in thermal expansion coefficients between the different phases induced additional tensile hydrostatic stresses. It must be emphasized that these three different contributions cannot be discriminated without in situ experiments, such as the ones presented here.

Similar behaviors were observed at a higher PT or lower QT; only experimental thermal expansion coefficients were reassessed for each experiment (2.575×10^{-5} K^{-1} for PT = 450 °C; 2.673×10^{-5} K^{-1} for QT = 200 °C). The respective evolutions of Δa_γ are represented in Figure 5b,c. For the sake of comparison, Table 1 summarizes the full variations in Δa_γ during steps (ii), (iv) and (vi) for the three tested conditions.

Table 1. Variations in Δa_γ for the three tested conditions during steps (ii) and (iv) and final cooling.

Variation in Δa_γ	Reference Cycle QT = 230 °C/PT = 400 °C	High Partitioning Temperature (PT) QT = 230 °C/PT = 450 °C	Low Quenching Temperature (QT) QT = 200 °C/PT = 400 °C
End of step (ii)	−0.0055 Å	−0.0060 Å	−0.0072 Å
During step (iv)	+0.0127 Å	+0.0106 Å	+0.0166 Å
During step (vi)	+0.0078 Å	+0.0087 Å	+0.0089 Å

Compressive states were observed after the initial martensitic transformation (step (ii)) for all studied cases. The lower the QT, the higher the pressure, as expected from Bruneseaux et al. [39]. Nevertheless, the measured differences were weak between experiments. During the partitioning step (iv), an increase was always observed, which could be attributed to carbon enrichment. For QT = 230 °C/PT = 450 °C and QT = 200 °C/PT = 400 °C cycles, the estimated carbon contents at the end of the partitioning step were about 0.67% and 0.80%, respectively. These estimates were again consistent with the *Ms* temperatures observed (or not) during final cooling.

4. Discussion

4.1. Carbon Mass Balances

The distribution of carbon between the austenitic and bcc phases was calculated by a direct carbon mass balance for the three studied Q&P conditions. The total carbon content in the bcc phases at RT was $C_0 - F\gamma \times C\gamma$, where $F\gamma$ was the final austenite fraction measured at RT, $C\gamma$ was the austenite carbon composition measured at PT (assumed to be unchanged at RT), and C_0 was the nominal carbon content. Table 2 summarizes these values for the three studied cycles.

Table 2. Carbon mass balances in austenite and in the bcc phases at RT after studied cycles. The phase fractions were deduced from Figures 3 and 4. The carbon contents in austenite were measured using the relative change in the lattice parameter in step (iv) given in Table 1.

Studied Cycles	Reference Cycle QT = 230 °C/PT = 400 °C	High PT QT = 230 °C/PT = 450 °C	Low QT QT = 200 °C/PT = 400 °C
$F\gamma$ (%)	27	30	21
$C\gamma$ (wt %)	0.70	0.67	0.80
$F\gamma \times C\gamma$ (wt %)	0.19	0.20	0.17
$C_0 - F\gamma \times C\gamma$ (wt %)	0.11	0.10	0.13

As already highlighted by HajyAkbari et al. [12] and Thomas et al. [15], carbon content remaining trapped in the bcc phases (martensite and bainite) was far from being negligible, which jeopardized the possibility of verifying strictly the original Speer's assumption (almost carbon-free bcc phases). The measured phase fractions are plotted in Figure 6 as functions of the QT and are compared to the calculations of Figure 2. The experimental data were higher than values expected by the model. Nevertheless, the effect of the QT seemed well captured: a lower QT led to a lower fraction of more stable austenite.

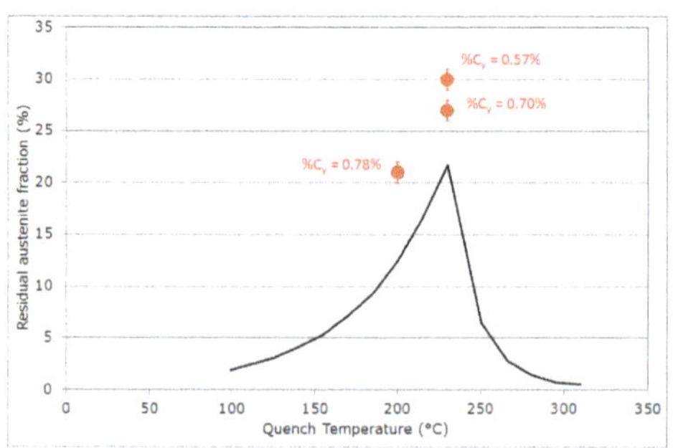

Figure 6. Retained austenite fractions at RT for the three studied conditions compared to the calculations presented in Figure 2. The error bars for phase fraction measurements are fixed to ±1%, as discussed in [23].

4.2. Critical Assessment of the Methods for Estimating Austenite Carbon Enrichment

The analyses presented above are based on an estimation of the carbon content derived from the experimental austenite lattice parameter measured at the end of the partitioning step. Using the procedure proposed by Toji et al. [18], with a lattice parameter value measured at RT at the end of the complete cycle, would have led to different conclusions, as illustrated in Table 3. This "post mortem" method led to significantly overestimating the carbon content in austenite for all the studied cycles; it did not allow accounting for the contribution of the hydrostatic internal stresses induced by the final martensitic transformation (as discussed by [18]) and thermal eigenstrains occurring during the final quench [42]. The carbon content in austenite obtained using the Toji approach was shown to be too high to explain the *Ms* temperatures observed during the final cooling, in contrast to what was observed using the lattice parameter value at the end of the partitioning step. Another consequence of this overestimation, highlighted in Table 3, was the resulting underestimation of the carbon content trapped in martensite (segregated or precipitated), which was still available for further carbide precipitation.

Table 3. Carbon mass balances in austenite and in the bcc phases at room temperature (RT) after the different studied cycles. The carbon contents in austenite were deduced from the measured austenite lattice parameter at RT using the Toji et al. formula [18].

Studied Cycles	Reference Cycle QT = 230 °C/PT = 400 °C	High PT QT = 230 °C/PT = 450 °C	Low QT QT = 200 °C/PT = 400 °C
a_γ (Å) at RT	3.6030	3.6013	3.6054
$C\gamma$ (wt % using the Toji et al. formula)	0.92	0.87	0.99
$C_0 - F\gamma \times C\gamma$ (wt %)	0.05	0.04	0.09

Nevertheless, this method, which is commonly used when only post-mortem XRD measurements [15,16,18] are available, grants access to relevant trends, as the hydrostatic internal stresses inherited from martensitic phase transformations and thermal eigenstrains are not very sensitive to processing conditions (please refer to Table 1).

5. Conclusions

In situ high-energy X-ray diffraction investigations of different Q&P cycles have demonstrated that:

- After the first martensitic transformation during cooling down to the QT, no significant evolution is observed during the re-heating step-up to the PT.
- During holding at the PT, an increase in the austenite lattice parameter is observed, resulting from the carbon redistribution from martensite and formation of ferritic bainite.
- A final increase of the austenite lattice parameter is observed during the final quench, attributed to internal stresses resulting from differences in the thermal expansion between the different phases present (and potentially improving the TRIP ability of the steel).

Due to this additional internal stress, the austenite carbon content cannot be estimated from its final lattice parameter, but from its value at the end of the partitioning step. A comparison of the carbon mass balances shows that using the RT lattice parameter leads to an underestimation, by a factor of 2, of the amount of carbon trapped in the bcc phases formed.

Acknowledgments: This work was supported by the French State through the project CAPNANO (ANR-14-CE07-0029) operated by the National Research Agency (ANR). The synchrotron experiments were realized in December 2014, under the MA2305 grant at the ESRF in Grenoble, which is fully acknowledged. The authors would like also to thank MATERIALIA cluster and the LABEX DAMAS (ANR-11-LABX-0008-01) from Lorraine for their support.

Author Contributions: Sébastien Yves Pierre Allain, Guillaume Geandier and Jean-Christophe Hell conceived and performed the experiments at ESRF; the data were analyzed with the help of Michel Soler, Mohamed Gouné and Frédéric Danoix, especially for the discussions about mass balances; Sébastien Yves Pierre Allain wrote the paper.

Conflicts of Interest: The authors declare no conflict of interest.

References

1. Barnard, S.J.; Smith, G.D.W.; Sarikaya, M.; Thomas, G. Carbon atom distribution in a dual phase steel: An atom probe study. *Scr. Metall.* **1981**, *15*, 387–392. [CrossRef]

2. Gouné, M.; Danoix, F.; Allain, S.; Bouaziz, O. Unambiguous carbon partitioning from martensite to austenite in Fe–C–Ni alloys during quenching and partitioning. *Scr. Mater.* **2013**, *68*, 1004–1007. [CrossRef]

3. Matas, S.; Hehemann, M.F. Retained Austenite and the Tempering of Martensite. *Nature* **1960**, *187*, 685–686. [CrossRef]

4. Speer, J.G.; Edmonds, D.V.; Rizzo, F.C.; Matlock, D.K. Partitioning of carbon from supersaturated plates of ferrite, with application to steel processing and fundamentals of the bainite transformation. *Curr. Opin. Solid State Mater. Sci.* **2004**, *8*, 219–237. [CrossRef]

5. Speer, J.; Matlock, D.K.; De Cooman, B.C.; Schroth, J.G. Carbon partitioning into austenite after martensite transformation. *Acta Mater.* **2003**, *51*, 2611–2622. [CrossRef]

6. Edmonds, D.V.; He, K.; Rizzo, F.C.; De Cooman, B.C.; Matlock, D.K.; Speer, J.G. Quenching and partitioning martensite—A novel steel heat treatment. *Mater. Sci. Eng. A* **2006**, *438–440*, 25–34. [CrossRef]

7. Somani, M.C.; Porter, D.A.; Karjalainen, L.P.; Misra, R.D.K. On Various Aspects of Decomposition of Austenite in a High-Silicon Steel During Quenching and Partitioning. *Metall. Mater. Trans. A* **2014**, *45*, 1247–1257. [CrossRef]

8. De Moor, E.; Speer, J.G.; Matlock, D.K.; Kwak, J.H.; Lee, S.B. Quenching and Partitioning of CMnSi Steels Containing Elevated Manganese Levels. *Steel Res. Int.* **2012**, *83*, 322–327. [CrossRef]

9. Xiong, X.C.; Chen, B.; Huang, M.X.; Wang, J.F.; Wang, L. The effect of morphology on the stability of retained austenite in a quenched and partitioned steel. *Scr. Mater.* **2013**, *68*, 321–324. [CrossRef]

10. Ariza, E.A.; Nishikawa, A.S.; Goldenstein, H.; Tschiptschin, A.P. Characterization and methodology for calculating the mechanical properties of a TRIP-steel submitted to hot stamping and quenching and partitioning (Q&P). *Mater. Sci. Eng. A* **2016**, *671*, 54–69. [CrossRef]

11. Zhao, H.S.; Li, W.; Zhu, X.; Lu, X.H.; Wang, L.; Zhou, S.; Jin, X.J. Analysis of the relationship between retained austenite locations and the deformation behavior of quenching and partitioning treated steels. *Mater. Sci. Eng. A* **2016**, *649*, 18–26. [CrossRef]

12. HajyAkbary, F.; Sietsma, J.; Miyamoto, G.; Furuhara, T.; Santofimia, M.J. Interaction of carbon partitioning, carbide precipitation and bainite formation during the Q&P process in a low C steel. *Acta Mater.* **2016**, *104*, 72–83. [CrossRef]

13. Pierce, D.T.; Coughlin, D.R.; Williamson, D.L.; Clarke, K.D.; Clarke, A.J.; Speer, J.G.; De Moor, E. Characterization of transition carbides in quench and partitioned steel microstructures by Mössbauer spectroscopy and complementary techniques. *Acta Mater.* **2015**, *90*, 417–430. [CrossRef]

14. Toji, Y.; Miyamoto, G.; Raabe, D. Carbon partitioning during quenching and partitioning heat treatment accompanied by carbide precipitation. *Acta Mater.* **2015**, *86*, 137–147. [CrossRef]

15. Thomas, G.A.; Danoix, F.; Speer, J.G.; Thompson, S.W.; Cuvilly, F. Carbon Atom Re-Distribution during Quenching and Partitioning. *ISIJ Int.* **2014**, *54*, 2900–2906. [CrossRef]

16. Clarke, A.J.; Speer, J.G.; Miller, M.K.; Hackenberg, R.E.; Edmonds, D.V.; Matlock, D.H.; Rizzo, F.C.; Clarke, K.D.; De Moor, E. Carbon partitioning to austenite from martensite or bainite during the quench and partition (Q&P) process: A critical assessment. *Acta Mater.* **2008**, *56*, 16. [CrossRef]

17. Santofimia, M.J.; Nguyen-Minh, T.; Zhao, L.; Petrov, R.; Sabirov, I.; Sietsma, J. New low carbon Q&P steels containing film-like intercritical ferrite. *Mater. Sci. Eng. A* **2010**, *527*, 6429–6439. [CrossRef]

18. Toji, Y.; Matsuda, H.; Herbig, M.; Choi, P.P.; Raabe, D. Atomic-scale analysis of carbon partitioning between martensite and austenite by atom probe tomography and correlative transmission electron microscopy. *Acta Mater.* **2014**, *65*, 215–228. [CrossRef]

19. Hell, J.C.; Dehmas, M.; Allain, S.; Prado, J.M.; Hazotte, A.; Chateau, J.P. Microstructure–Properties Relationships in Carbide-free Bainitic Steels. *ISIJ Int.* **2011**, *51*, 1724–1732. [CrossRef]

20. Hell, J.C. Aciers Bainitiques Sans Carbure: Caractérisations Microstructurale Multi-Echelle et In Situ de la Transformation Austénite-Bainite et Relations Entre Microstructure et Comportement Mécanique. Ph.D. Thesis, Université Paul Verlaine, Metz, France, 10 November 2011.

21. Caballero, F.G.; Allain, S.; Cornide, J.; Puerta Velásquez, J.D.; Garcia-Mateo, C.; Miller, M.K. Design of cold rolled and continuous annealed carbide-free bainitic steels for automotive application. *Mater. Des.* **2013**, *49*, 667–680. [CrossRef]

22. Nishikawa, L.; Ogata, P.; Nishikawa, A.; Ramirez, M.; Goldenstein, H. Tempering Behaviour of a Quenched Microalloyed Pipeline Steel. In Proceedings of the TMS 2016 Conference, Nashville, TN, USA, 14–18 February 2016.

23. Allain, S.Y.P.; Geandier, G.; Hell, J.C.; Soler, M.; Danoix, F.; Gouné, M. In-situ investigation of quenching and partitioning by High Energy X-Ray Diffraction experiments. *Scr. Mater.* **2017**, *131*, 15–18. [CrossRef]
24. Epp, J.; Hirsch, T.; Curfs, C. In situ X-Ray Diffraction Analysis of Carbon Partitioning during Quenching of Low Carbon Steel. *Metall. Mater. Trans. A* **2012**, *43*, 2210–2217. [CrossRef]
25. Van Bohemen, S.M.C. Modeling Start Curves of Bainite Formation. *Metall. Mater. Trans. A* **2010**, *41*, 285–296. [CrossRef]
26. Van Bohemen, S.M.C. Bainite and martensite start temperature calculated with exponential carbon dependence. *Mater. Sci. Technol.* **2012**, *28*, 487–495. [CrossRef]
27. The FIT2D Home Page. Available online: http://www.esrf.eu/computing/scientific/FIT2D/ (accessed on 11 April 2017).
28. Rodriguez-Carvajal, J. Recent advances in magnetic structure determination by neutron powder diffraction. *Physica B* **1993**, *192*, 55. [CrossRef]
29. Rementeria, R.; Jimenez, J.A.; Allain, S.Y.P.; Geandier, G.; Poplawsky, J.D.; Guo, W.; Urones-Garrote, E.; Garcia-Mateo, C.; Caballero, F.G. Quantitative assessment of carbon allocation anomalies in low temperature bainite. *Acta Mater.* **2017**, *133*, 333–345. [CrossRef]
30. Hirotsu, Y.; Nagakura, S. Crystal structure and morphology of the carbide precipitated from martensitic high carbon steel during the first stage of tempering. *Acta Metall.* **1972**, *20*, 846. [CrossRef]
31. Van Bohemen, S.M.C.; Santofimia, M.J.; Sietsma, J. Experimental evidence for bainite formation below M_s in Fe–0.66C. *Scr. Mater.* **2008**, *58*, 488–491. [CrossRef]
32. Kolmskog, P.; Borgenstam, A.; Hillert, M.; Hedstrom, P.; Babu, S.S.; Terasaki, H.; Komizo, Y.I. Direct Observation that Bainite can Grow Below M_s. *Metall. Mater. Trans. A* **2012**, *43A*, 4984–4988. [CrossRef]
33. Kim, D.; Lee, S.J.; De Cooman, B.C. Microstructure of Low C Steel Isothermally Transformed in the M_s to M_f Temperature Range. *Metall. Mater. Trans. A* **2012**, *43A*, 4967–4983. [CrossRef]
34. Navarro-López, A.; Hidalgo, J.; Sietsma, J.; Santofimia, M.J. Characterization of bainitic/martensitic structures formed in isothermal treatments below the M_s temperature. *Mater. Charact.* **2017**, *128*, 248–256. [CrossRef]
35. Bigg, T.D.; Edmonds, D.V.; Eardley, E.S. Real-time structural analysis of quenching and partitioning (Q&P) in an experimental martensitic steel. *J. Alloys Compd.* **2013**, *577*, 695–698. [CrossRef]
36. Lu, X.G.; Selleby, M.; Sundman, B. Assessments of molar volume and thermal expansion for selected bcc, fcc and hcp metallic elements. *Calphad* **2005**, *29*, 68–89. [CrossRef]
37. Onink, M.; Brakrnan, C.M.; Tichelaar, F.D.; Mittemeijer, E.J.; Van der Zwaag, S. The lattice parameters of austenite and ferrite in FeC alloys as functions of carbon concentration and temperature. *Scr. Metall. Mater.* **1993**, *29*, 1011–1016. [CrossRef]
38. Van Bohemen, S.C.M. The nonlinear lattice expansion of iron alloys in the range 100–1600 K. *Scr. Mater.* **2013**, *69*, 315–318. [CrossRef]
39. Bruneseaux, F. Apport de la Diffraction des Rayons X à Haute Energie sur les Transformations de Phases, Application aux Alliages de Titanes. Ph.D. Thesis, The National Polytechnic Institute of Lorraine (INPL), Nancy, France, 16 May 2008.
40. Ghosh, G.; Olson, G.B. The isotropic shear modulus of multicomponent Fe-base solid solutions. *Acta. Mater.* **2002**, *50*, 2655–2675. [CrossRef]
41. Manjoine, M.J.; Voorhees, H.R. *Compilation of Stress-Relaxation Data for Engineering Alloy*; ASTM Data Series Publication DS 60; ASTM International: West Conshohocken, PA, USA, 1982.
42. Arsenault, R.J.; Taya, M. Thermal residual stress in metal matrix composite. *Acta Metall.* **1987**, *35*, 651–659. [CrossRef]
43. Lu, P. Further studies on Mori–Tanaka models for thermal expansion coefficients of composites. *Polymer* **2013**, *54*, 1691–1699. [CrossRef]

Article

In-Situ Investigation of Strain-Induced Martensitic Transformation Kinetics in an Austenitic Stainless Steel by Inductive Measurements

Carola Celada-Casero [1,2,*], Harm Kooiker [3,4], Manso Groen [3], Jan Post [3,5] and David San-Martin [1]

[1] Materalia Research Group, Centro Nacional de Investigaciones Metalúrgicas (CENIM-CSIC),
Av. Gregorio del Amo 8, 28040 Madrid, Spain; dsm@cenim.csic.es
[2] Department of Materials Science and Engineering, Delft University of Technology, Mekelweg 2,
2628 CD Delft, The Netherlands
[3] Philips Advanced Technology Centre, P.O. Box 20100, Tussendiepen 4, 9200 CA Drachten, The Netherlands;
Harm.Kooiker@philips.com (H.K.); Manso.Groen@philips.com (M.G.);
J.Post@philips.com or Jan.Post@rug.nl (J.P.)
[4] Department of Nonlinear Solid Mechanics, University of Twente, Drienerlolaan 5, 7522 NB Enschede,
The Netherlands
[5] Faculty of Science and Engineering, Advanced Production Engineering—Engineering and Technology
Institute Groningen, Nijenborgh 4, 9747 AG Groningen, The Netherlands
* Correspondence: C.CeladaCasero@tudelft.nl; Tel.: +31-015-27-82249

Received: 31 May 2017; Accepted: 12 July 2017; Published: 13 July 2017

Abstract: An inductive sensor developed by Philips ATC has been used to study in-situ the austenite (γ) to martensite (α') phase transformation kinetics during tensile testing in an AISI 301 austenitic stainless steel. A correlation between the sensor output signal and the volume fraction of α'-martensite has been found by comparing the results to the ex-situ characterization by magnetization measurements, light optical microscopy, and X-ray diffraction. The sensor has allowed for the observation of the stepwise transformation behavior, a not-well-understood phenomena that takes place in large regions of the bulk material and that so far had only been observed by synchrotron X-ray diffraction.

Keywords: stainless steel; metastable austenite; strain-induced martensite; transformation kinetics; inductive measurements

---

1. Introduction

One of the key features of the metastable austenitic stainless steels (MASSs) is the Transformation Induced Plasticity (TRIP) effect [1]. In their annealed soft state, MASSs are fully austenitic (γ); however, they can be strengthened by grain refinement [2] and by its transformation into martensite (α') under the application of stress/strain [3,4]. Depending on their composition, MASSs can be further strengthened by an aging treatment that induces the precipitation of strengthening particles in the martensite phase, as it is the case of maraging and precipitation hardening (PH) stainless steels [5].

Strain-induced martensite forms from austenite when this is subjected to deformation above its flow stress. When the plastic deformation stops, the strain-induced martensitic transformation also stops. However, depending on the residual stress level in the material, the transformation may also take place immediately after plastic deformation, which is known as stress-assisted transformation [6]. The occurrence of the $\gamma \rightarrow \alpha'$ phase transformation is closely linked to how plastic deformation takes place in the austenite phase, which is connected to the stacking fault energy (SFE) of the steel. In alloys with SFE lower than 15 mJ/m^2, the martensitic transformation is favoured and, thus, TRIP occurs. On the other hand, in alloys with a SFE between 25 and 60 mJ/m^2, the austenite deforms by twinning and the deformation is mediated by dislocation slip [7]. Typically, MASSs have low SFEs (about 20 mJ/m^2) and, consequently, the deformation of the austenite is initially characterized by the formation of staking faults (SFs). Eventually, larger deformations give rise to shear bands and planar defects that result from the overlapping of SFs on austenite {111} planes and which are well known as preferred sites for the nucleation of ε-martensite, a precursor of α'-martensite [4,7–12]. As reported in the literature, the occurrence of the martensitic transformation via the formation of the intermediate ε-martensite phase ($\gamma \rightarrow \varepsilon \rightarrow \alpha'$) or directly ($\gamma \rightarrow \alpha'$), as well as the transformation rate, depend on austenite stability. A number of factors, such as the chemistry, grain size, temperature, and the strain rate, influence austenite stability [3,4,6–14]. The formation of martensite during the application of stresses and strains influences significantly the response of the material during processes like sheet metal forming. The increase of the work-hardening of the steel as a consequence of the martensitic transformation results in enhanced strength and formability, which makes MASSs good candidates for multi-stage processes [9,10,15]. However, using a number of steps to obtain the desired final shape is an extra complication. An accurate determination of the martensite transformation kinetics as a function of the plastic strain is, therefore, necessary to improve the design and prediction of metallurgical processes.

It is not easy to study the evolution of the martensitic transformation under straining. In-situ investigations generally requires the use of large facilities and complex techniques as high energy synchrotron X-ray diffraction (HEXRD) or neutron diffraction, where allocation of beamtime is not straightforward [8,9,11,16]. For this reason, the quantification of martensite by metallography, magnetization measurements or diffraction methods on post-mortem tensile-tested specimens is usually preferred. However, it requires a great number of experiments and long times of experimental procedure [14]. Besides, this method does not provide information on possible stress-assisted transformation induced by residual stresses in the material right after plastic straining. This study presents a method based on an inductive sensor developed by Philips ATC (Drachten, The Netherlands) that allows for a simple and fast in-situ characterization of the formation kinetics of mechanically-induced martensite during tensile testing. The procedure is validated and the results are compared to those obtained by the currently used ex-situ methods.

2. Materials and Methods

The material investigated in this work is an austenitic metastable stainless steel, whose chemical composition is given in Table 1 and corresponds to the AISI 301 grade. The material was supplied as cold rolled sheets of 0.30 mm in thickness in the full annealed state.

Table 1. Chemical composition of AISI 301 grade (in wt %) with balanced Fe.

C	Cr	Ni	Si	Mn	Mo
Max 0.15	16–18	6.5–9	<1.5	<2	<0.8

The mechanical behavior of the annealed material was studied at room temperature by means of uniaxial tensile tests. Standard specimens were machined perpendicularly to the rolling direction according to the ASTM E8/E8M-09 standard and tested in a Zwick-Roell-Z030 universal testing machine (GmbH & Co. KG, Ulm, Germany) at a strain rate of $\dot{\varepsilon} = 5 \times 10^{-4}$ s^{-1}. The strain was measured using double-sided clip extensometers with a parallel length of 80 mm.

The martensite transformation behavior was studied in-situ during tensile testing by means of an inductive sensor developed by Philips ATC that is able to measure differences in relative magnetic permeability ($\mu_r \approx 100$) between the paramagnetic austenite and the ferromagnetic martensite phases. A standard integrated circuit (IC) is used for the electronics. An alternating current (AC) is sent through two coils connected in series, i.e., the measurement coil and the reference coil, and the difference in voltage across them is measured and amplified. The coils are 31 mm in length. A number of adjusting elements can be used to set the voltage, current, frequency, and amplification factor. The IC contains an oscillator that generates voltage, a filter, and an amplifier. A frequency of 7.5 kHz was selected in order to avoid as much external electromagnetic interference as possible [15]. The setup is schematically illustrated in Figure 1. The sensor is placed in the middle of the tensile specimen's gage length and the output signal is recorded during the course of the mechanical testing. The correlation between the volume fraction of martensite and the voltage signal from the sensor is, however, not trivial and requires the correction of some effects influencing the signal such as applied stress, plastic strain, temperature, and specimen volume inside the coil. Of great importance is the correction of the stress-effect. For this purpose, cyclic loading/unloading/reloading (LUR) tensile tests were performed and the inductive signal was registered by using different currents circulating through the coil. The specimens were unloaded to zero force and reloaded again in applied strain increments of 0.05 until the samples fractured. The comparison of the output signal between the loaded and unloaded states allows for the correction of the stress effect. Once the signal is cleaned, a direct relationship between the content of martensite and the sensor output signal can be established based on ex-situ characterization of the martensite volume fraction.

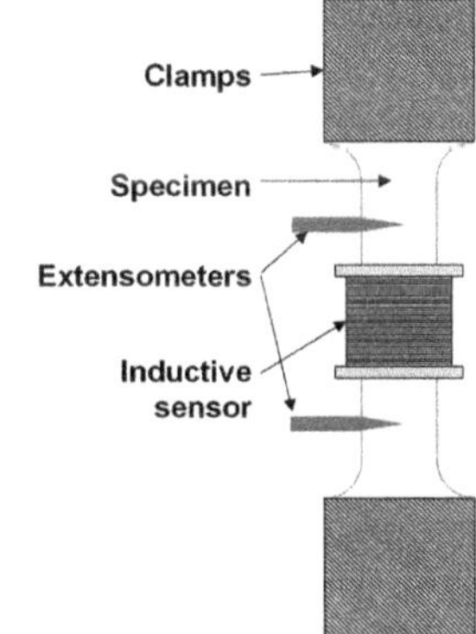

Figure 1. Tensile test setup with inductive sensor.

Interrupted tensile tests were performed in applied strain increments of 0.05 between 0 and the fracture ($\varepsilon_E^{\,F} = 0.52$). The post-mortem specimens were used to ex-situ determine the volume fraction of martensite by different characterization techniques, as explained below.

The volume fraction of martensite was measured ex-situ by different characterization techniques: light optical microscopy (LOM), X-ray diffraction (XRD), and magnetization measurements. For LOM characterization, a Nikon Epiphot 200 (Nikon Instruments, Inc., Melville, NY, USA) was used. The samples were cold mounted and the cross sections were prepared following the standard procedure: grinding down to 1200 grit using different grit size sandpapers and polishing in two steps using different lap cloths and finishing with 1 µm diamond solution. The microstructure was revealed by color etching using HCl-base Beraha's reagent for 10 s. This etching provides a good contrast between martensite (dark) and austenite (light). By using "ImageJ" image analysis software (version 1.47s, NIH, Stapleton, NY, USA, 2013), the LOM micrographs were converted to binary images. By thresholding the greyscale intensity, the dark areas (martensite phase) are included in the area of analysis and its area fraction can be determined. Both automatic and manual thresholds were compared for evaluation of the method. A total area of 300×250 µm^2 was analyzed for each condition.

A quantum design MPMS-XL SQUID magnetometer (Quantum Design, Inc., San Diego, CA, USA) was employed to measure the magnetization curve at room temperature. Square-shaped samples of 3 mm side were extracted from the middle of the gage length and the magnetization was recorded by varying the external applied magnetic field from 0 to 5 T in steps of 0.2 T. The volume fraction of martensite correlates with the magnetization saturation of the samples (M_{sat}), through $f_{\alpha'} = M_{sat} / M_{sat}{}^{\alpha'}$, as explained elsewhere [17]. To obtain a reference magnetization saturation ($M_{sat}{}^{\alpha'}$), a piece of as-received material was severely cold-rolled by reducing the strip thickness more than 60% in order to induce the complete transformation into martensite. A M_{sat} of 133 Am^2/kg was obtained.

Specimens of 1 cm^2 were prepared for XRD as explained for LOM and including a final polish with colloidal silica suspension (40 nm). XRD measurements were carried out with a Bruker AXS D8 diffractometer (Bruker AXDS GmbH, Karlsruhe, Germany) equipped with a Co-X-ray tube, Goebel mirror optics collimator and a LynxEye Linear Position Sensitive Detector for ultra-fast XRD measurements. The area of interest on the sample was accurately and precisely located in the X-ray beam by using a laser/video microscope from Bruker AXS. A current of 30 mA and a voltage of 40 kV were employed as tube settings. The XRD spectra were collected over a 2θ range of 35–135° with a step size of 0.01°. The Rietveld analysis program TOPAS (version 4.2, Bruker AXS, 2009) was used for the quantification of the α'-martensite and γ-austenite phases and the calculation of their structural parameters. Crystallographic information of the different phases for the Rietveld refinement was obtained from Pearson's Crystal Structure Database for Inorganic Compounds. The refinement protocol also included other parameters like background, zero displacement, the scale factors, the peak breadth, and the unit cell parameter. In order to eliminate the instrumental contribution to peak broadening, instrument functions were empirically parameterized from the profile shape analysis of a corundum sample measured under the same conditions.

3. Results

3.1. Ex-Situ Characterizatiton of Strain-Induced α'-Martensite Formation

The progress of the α'-martensite formation with the strain was studied by LOM, magnetization measurements, and XRD in tensile-tested specimens interrupted at different strains (ε_E). Figure 2 displays the X-ray diffractograms obtained for the as-received, initial microstructure (IM), post-mortem tensile-tested specimens strained to ε_E = 0.10, 0.20, 0.30, and 0.40 and a strained-to-fracture specimen (ε_E = 0.52). The initial microstructure is fully austenitic. As the material is strained, the peaks related to the γ-austenite phase decrease and those related to α'-martensite increase. It is worth mentioning that the presence of the ε-martensite phase is detected at strains of 0.20 or higher, as highlighted with diamonds in the diffractograms. These peaks, nevertheless, are quite weak. A maximum ε-martensite volume fraction of 0.04 was detected for the condition strained to ε_E = 0.30. This might be indicating that, for stains lower than 0.20, the amount of ε-martensite is too small to be detected.

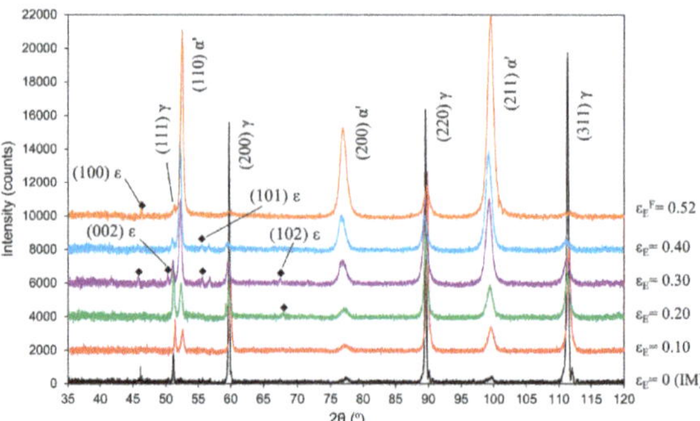

Figure 2. X-ray diffractograms for the initial microstructure (IM) and interrupted tensile-tested specimens at strains (ε_E) of 0.10, 0.20, 0.30, 0.40, and fracture (ε_E = 0.52). Diamond dots indicate the presence of peaks associated to ε-martensite phase.

Figure 3 shows LOM micrographs of samples strained to 0.10, 0.20, 0.30, and 0.40. The martensite phase is revealed in dark colors and the austenite is unetched. After the application of small strains ($\varepsilon_E \leq 0.10$), parallel slip bands appeared in those grains that are favorably oriented with respect to the applied stress direction, where slip bands are localized crystallographic slip and, thus, cannot cross existing interfaces into other orientations [18]. By increasing the strain, shear bands appear when homogeneous dislocation slip is inhibited or when an insufficient number of crystallographic slip systems is available in narrow deformation zones (order of microns) while the rest of the matrix undergoes comparably low and homogeneous plastic flow [18]. As the deformation proceeds, the shear bands intersect one another [12,19]. These intersections act as nucleation sites for α'-martensite embryos, which coalesce and grow parallel to the shear bands, as observed for samples strained to $\varepsilon_E = 0.10$ and $\varepsilon_E = 0.20$. With larger strains, the fraction of α'-martensite increases rapidly. Since the purpose of etching is the quantification of the martensite phase, other than revealing the microstructure, it is important to mention here that slip bands are also revealed with the etching. This fact directly leads to inaccuracy of the method. The quantification procedure is commented in the following paragraph.

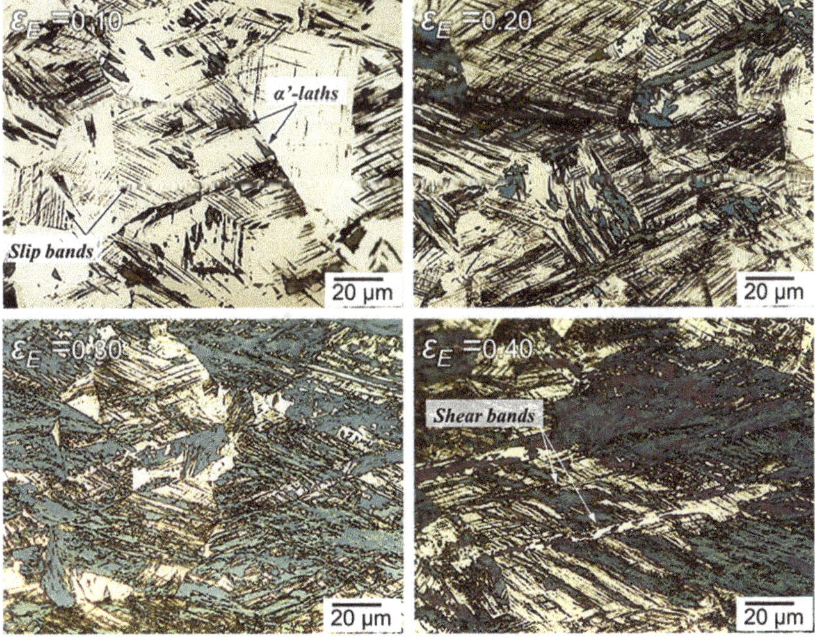

Figure 3. Light-optical micrographs of tensile-tested specimens interrupted at strains (ε_E) of 0.10, 0.20, 0.30, and 0.40 after etching with Beraha's reagent for 10 s. The martensite phase is revealed in black and blue, whereas the austenite phase remains unetched.

Figure 4a displays a black-and-white image for the sample strained to $\varepsilon_E = 0.10$ (Figure 3) after conversion to a binary image and application of a grey scale threshold value (λ_{th}) of 127. The martensite phase and slip bands correspond to black pixels, while the white ones correspond to the austenite phase. Figure 4b shows the histogram of the grey scale image for the same condition. It was found that the estimated martensite content depends strongly on the choice of λ_{th}. Threshold value variations of ± 5 around $\lambda_{th} = 127$ results in changes of ± 0.01 in volume fraction of martensite. Besides, the grey scale histogram also depends on the etching time. A standard deviation value was estimated considering the automatic λ_{th} detected by the software and a subjectively chosen λ_{th} value that gives the best representation of the etched phase.

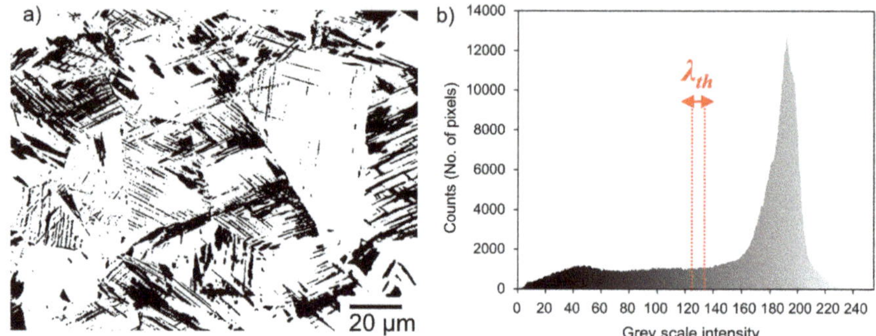

Figure 4. (**a**) Black-and-white image obtained from the light-optical micrograph of the condition strained up to $\varepsilon_E = 0.10$ after applying a grey scale threshold value $\lambda_{th} = 127$; (**b**) Histogram of the same micrograph showing the number of pixels vs. the grey scale intensity value.

Figure 5 displays the evolution of the α'-martensite volume fraction with the engineering strain obtained from post-mortem tensile-tested by metallographic quantification, magnetization measurements, and XRD. A fairly good agreement is obtained between the data from magnetization and XRD. Since the ε-martensite is a paramagnetic phase, its formation cannot be detected by magnetization measurements. Differences between XRD and magnetization must be due to experimental errors and to the fact that magnetization measurements is a bulk characterization technique, whereas XRD is restricted to the surface. Besides, the discrepancies observed for the initial microstructure ($\varepsilon_E = 0$) could be due to some martensite induced during the metallographic preparation required for both XRD and optical inspection. The values obtained from metallographic quantification appear to be overestimated. The presence of ε-martensite in the microstructure is negligible, compared to that of γ-austenite or α'-martensite. However, it is revealed dark after etching, as are slip and shear bands. By metallographic quantification, these features are accounted for as martensite phase and, thus, this could be a possible reason for the overestimation. The relative error of the data from magnetization measurements is smaller than 0.5%. In the case of XRD, it is difficult to quantify the total error as it depends on the calculation method, instrumental limitation and the sample condition. Differences between XRD and magnetization must be due to experimental errors and to the fact that magnetization measurements is a bulk characterization technique, whereas XRD is restricted to the surface. Although both the result from magnetization and XRD show consistency to one another, measuring the bulk magnetization is a more effective and representative technique to determine the volume fraction of α'-martensite.

The volume fraction of strain-induced α'-martensite ($f_{\alpha'}$) can be effectively described by the kinematic equation [20]:

$$f_{\alpha'} = f_s \left[1 - \exp\left(-\beta \varepsilon_{pl}{}^n \right) \right] \tag{1}$$

where f_s is the saturation value of transformed martensite, ε_{pl} is the plastic strain, and the parameters β and n represent the stability of the austenite phase and the deformation mode, respectively. In this work, $f_s = 0.92$ and corresponds to the fraction of α'-martensite magnetically measured in a specimen strained to fracture. The fitting parameters were found to be $\beta = 13.4$ and $n = 2.2$ for the data obtained by magnetization measurements. They also represent well the experimental data from XRD and are consistent with other investigations in similar steels tensile strained at room temperature [20].

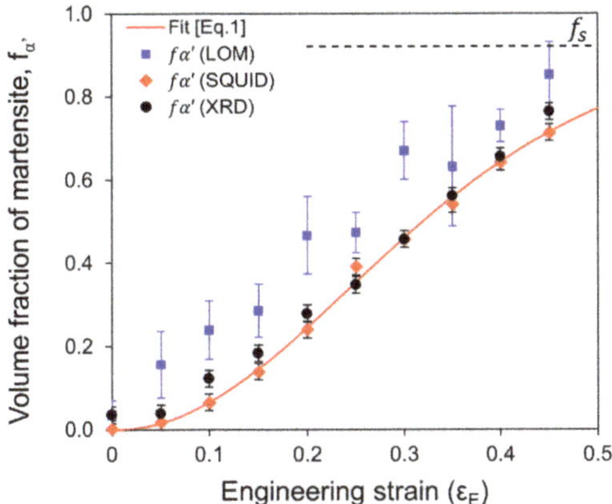

Figure 5. Evolution of martensite volume fraction ($f_{\alpha\prime}$) with strain (ε_E) obtained ex-situ from interrupted tensile tests determined by LOM, magnetization measurements (SQUID), and XRD. The dashed line represents the saturation value of transformed martensite (f_s = 0.92) found by XRD for a sample strained to fracture.

3.2. In-Situ Inductive Measurements

To transform the inductive signal from the sensor into volume fraction of martensite, the various factors affecting it have to be separated: (1) the material volume inside the sensor; (2) the temperature; (3) the applied stress and (4) the volume fraction of martensite. In this section, the different influences are decoupled and the inductive signal is corrected to isolate the influence of the martensite volume fraction.

Since the sensor measures the μ_r of the enclosed material volume, the output signal will be affected by any change in this volume. As the material is elongated during the tensile test, the cross sectional area of the specimen decreases accordingly and, in the same manner, the volume enclosed within the sensor. Therefore, the voltage signal decreases linearly during tensile testing as the volume enclosed within the coil is reduced. In order to make the measurements comparable, the voltage values have to be recalculated for the reference volume, where the reference volume refers to the maximum volume of sample inside the sensor. This situation corresponds to the initial state, before straining. The volume effect on the voltage values is then corrected through the following expression: $V_V = V_{V0} \cdot (1 + \Delta l / L_0)$, where V_{V0} stands for the voltage signal obtained with the maximum volume of sample inside the sensor, corresponding to the non-deformed initial state, and $\Delta l / L_0$ is the strain increment measured by the extensometers, which accounts for the reduction in volume within the sensor if isotropic behavior is assumed.

Both the self-induction of the coil and the μ_r of the material are temperature dependent. The coil temperature tends to increase as the result of the current circulating through it. Also, the specimen is warmed up by the deformation if high strain rates are used and thus the chemical driving force for the transformation ($\Delta G^{\gamma \to \alpha\prime}$) decreases [4,19,21]. Since the strain rate used in this work ($\dot{\varepsilon} = 5 \times 10^{-4}\,\mathrm{s}^{-1}$) does not induce heating of the specimen during the deformation [13], the current circulating through the coil is the only heat source. It is known that magnetic fields above 1 T are required to magnetically saturate the sample and, thus, to obtain a magnetization response that can be reliably related to the whole volume fraction of ferromagnetic phase [17,22]. However, the application of magnetic fields larger than 1 T for several minutes induce the formation of a significant volume fraction of isothermal martensite [23]. In this work, a compromise is reached by setting the current to 55 mA, which generates a constant magnetic field of 0.6 T. This magnetic field is high enough to capture martensite formation, but sufficiently low to minimize the temperature increment within the measurement time. Besides, this magnetic field will not accelerate the martensitic transformation kinetics during the course of the cyclic load/unload experiments, which take around 15 min.

The magnetomechanical or "Villari" effect is the phenomenon that relates the effect of an applied stress on the magnetization response of a ferromagnetic material. Thus, it depends on the fraction of martensite and the hydrostatic stress [6]. Cyclic LUR experiments give information about the change of magnetic properties driven by the elastic volume changes in the stressed material. Figure 6a shows the coupled influence of the plastic strain and current and, thus, the magnetic field on the voltage recorded. On the one hand, it is observed that the higher the current, the higher the output voltage signal and thus, the higher magnetic field. On the other hand, in each unload/reload cycle, the voltage first increases and then decreases significantly, reaching a minimum when the load is zero. Then the specimen is reloaded and the voltage raises again. The resultant voltage loop induced by the unloading-reloading cycles is a consequence of the changes in the magnetic properties, and becomes more pronounced as the current is increased. By setting the current to 55 mA (red curve of Figure 6a), the stress effect is significantly reduced. The lower the current, the smaller the change in voltage due to unloading and reloading, as observed from the $V - \varepsilon_E$ curves in Figure 6a. It is worth mentioning that it is not possible to eliminate the stress effect completely. A further decrease of the current would not generate a sufficiently high magnetic field to magnetically saturate the sample, as discussed previously. Figure 6b displays the LUR experiment with the $V - \varepsilon_E$ curve obtained with a current of 55 mA. It is seen that the stress effect increases as the sample is strained and, thus, as the fraction of martensite increases. Figure 6c gives a magnified view of Figure 6b for ε_E in the range 0.45–0.50. In this view, the hysteresis behavior can be better appreciated in the stress–strain curve upon unloading/reloading and in terms of its influence on the inductive signal. Under zero stress, the voltage signal is at its minimum and it can be said that this value corresponds solely to the martensitic volume fraction as there is no stress-effect. The determination of this value free of stress-effect is denoted as $V_{\sigma 0}$ (pointed by arrows in Figure 6c) and has a systematic error of ± 0.01. V_{σ} refers to the signal obtained in the loaded condition and, thus, it is the sum of the martensite fraction and the stress effect. Therefore, the effect of the stress can be isolated by subtracting $V_{\sigma 0}$ from V_{σ}. Figure 6d shows the evolution of both signals V_{σ} (loaded) and $V_{\sigma 0}$ (unloaded) with the strain and, thus, the deconvolution of two effects. The plastic deformation is also known to affect the magnetic response of ferromagnetic materials due to texture changes and the effect of dislocations on the domain structure and domain walls movement [24]. However, the texture change of α'-martensite during tensile testing is small [25] and, therefore, the difference between $V_{\sigma 0}$ and V_{σ} can be solely attributed to the magnetomechanical effect. In order to find a function that allows for the correction of the output signal obtained under monotonic tensile tests, the elastic stress effect is expressed as the ratio $V_{\sigma}/V_{\sigma 0}$ and plot in the secondary y-axis of Figure 6d. It is observed that, for the applied current, there is no stress influence until a plastic strain threshold value of $\varepsilon_E^{th} = 0.31$ is exceeded. Afterwards, the stress effect increases linearly with the plastic strain and the output signal can be corrected through the following expression: $V_{\sigma}/V_{\sigma 0} = 0.8 + 0.7\varepsilon_E$, where the correction factor is a function of the current circulating through the coil.

After applying the above mentioned corrections, the inductive signal $V_{\sigma 0}$ is solely due to the martensite volume fraction. In order to find out the correlation between the voltage and the fraction of martensite phase, the results from SQUID magnetization measurements were compared with the inductive values ($V_{\sigma 0}$) corresponding to the same strain level. Figure 7 shows a plot of $V_{\sigma 0}$ as a function of the volume fraction of α'-martensite obtained by SQUID magnetization measurements. Y-error bars are of the order of data points. The exponential fit equation utilized to correlate both signals will be used to convert the voltage signal into volume fraction of α'-martensite obtained during the monotonic tensile tests.

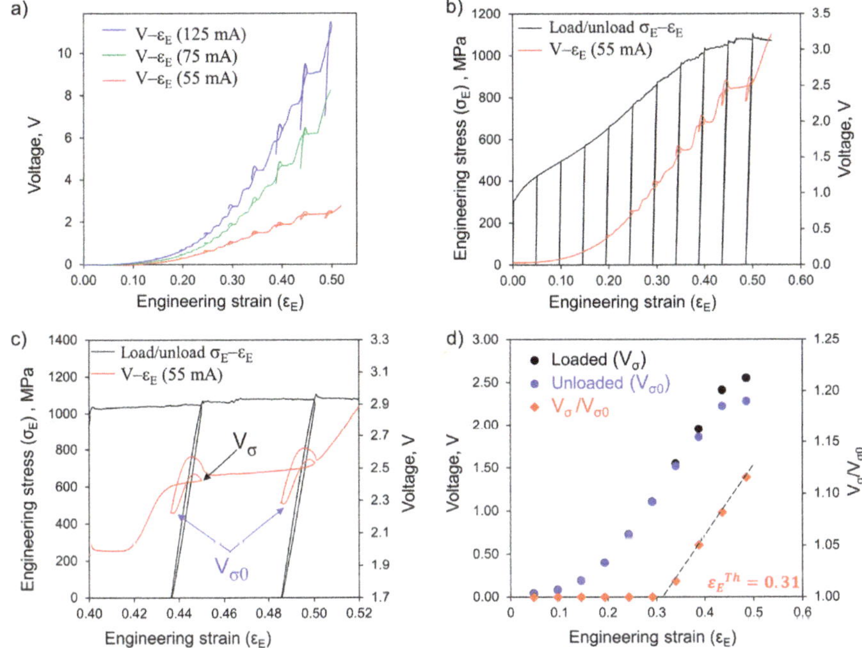

Figure 6. (**a**) Evolution of inductive signals with engineering strain (ε_E) obtained during the course of cyclic load/unload tests by using currents of 125, 75, and 55 mA. (**b**) Load–unload σ_E–ε_E curve and inductive signal obtained with a current of 55 mA. (**c**) Magnification of (**b**) for the range ε_E = 0.40–0.52. $V_{\sigma 0}$ and V_σ stand for voltage signal at zero stress and under stress, respectively. (**d**) Evolution of both the voltage signal in the loaded (V_σ) and the unloaded ($V_{\sigma 0}$) states recorded during the load-unload cyclic measurements.

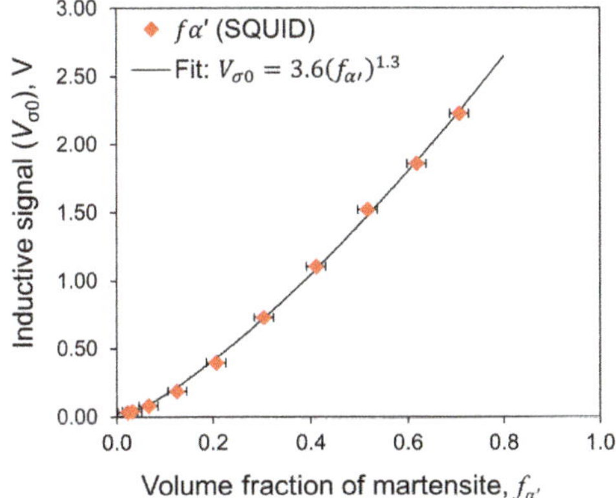

Figure 7. Relationship between the signal output (voltage) and the martensite content estimated using SQUID magnetization measurements.

Figure 8 displays the σ_E–ε_E curve obtained under monotonic tensile test to fracture. The material exhibits a yield strength and ultimate tensile strength of 310 and 1060 MPa, respectively. Both the uniform and total elongations are found at an engineering strain of 0.50. A similar evolution of $f_{\alpha\prime}$ with

ε_E is obtained by SQUID magnetization measurements, the sensor, and the formation kinetics predicted by Equation (1). The data obtained inductively provides, nevertheless, more information about processes occurring during the martensitic transformation. Figure 9a represents the true stress–strain curves ($\sigma_T - \varepsilon_T$) and the evolution of the martensite formation as a function of ε_T. A transition between a smooth transformation and a "step-like" transformation behavior is observed from strain levels of $\varepsilon_T \approx 0.23$ ($f_{\alpha'} \approx 0.32$) onwards. Figure 9b provides a detailed view of this stepwise behavior. In each transformation step (pointed by arrows), a large fraction of austenite suddenly transforms into α'-martensite in a narrow strain increment. Each burst of martensite is followed by a plateau where almost no transformation occurs. These sudden increments in martensite volume fraction are reflected as roughness in the $\sigma_E - \varepsilon_E$ curve.

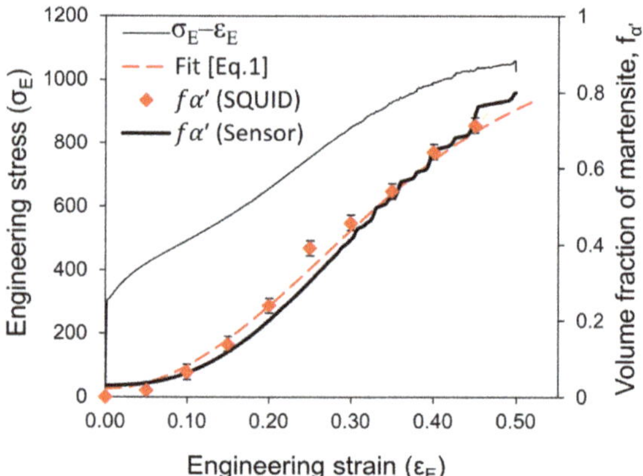

Figure 8. Engineering stress–strain ($\sigma_E - \varepsilon_E$) curve and evolution of the α'-martensite volume fraction ($f_{\alpha'}$) with ε_E obtained by ex-situ magnetization measurements (SQUID), in-situ inductive measurements (sensor) and as predicted by the kinetics model.

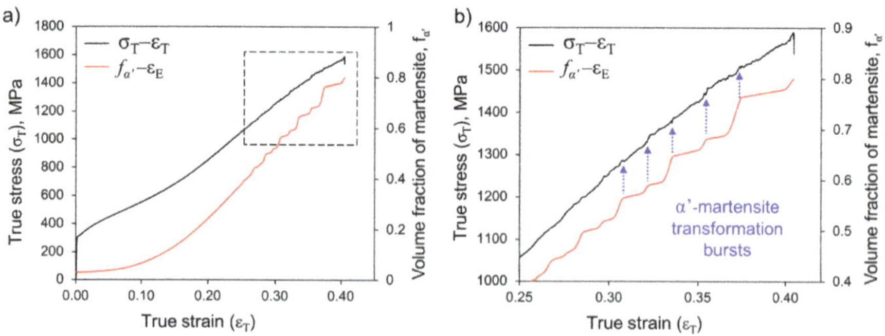

Figure 9. (**a**) True stress–strain ($\sigma_T - \varepsilon_T$) curve and evolution of $f_{\alpha'}$ with ε_T obtained in-situ by using the sensor. (**b**) Magnification of the region marked with a square in (**a**) pointing out the effect of the transformation bursts on the strengthening of the material.

4. Discussion

4.1. Comparing Techniques to Desctibe the Martensitic Transformation Kinetics under Unixial Tension

The quantification of the α'-martensite by the different ex-situ characterization techniques is compared in Figure 5. Metallographic quantification leads to an overestimation of α'-martensite volume fraction as compared to the results from magnetization measurements and XRD. The reason is most likely related to the fact that etching does not only disclose α'-martensite, but also ε–martensite, and slip and shear bands. Other than that, the results are affected by the selected λ_{th} value and the etching of the sample, which makes the method subjective and ambiguous. Therefore, this technique is not recommended for determining the volume fraction of martensite. On the contrary, magnetization measurements and XRD represent more accurate and reliable quantification techniques [22]. Both methods show consistency to one another and also with the results obtained in-situ with the sensor (Figure 8).

It is an important drawback of ex-situ characterization techniques that none of them are able to detect the stress-assisted type of martensitic transformation. Post-mortem tensile specimens are likely to contain residual stresses accumulated in the microstructure after straining. These stresses might trigger further martensitic transformation in the meantime between tensile testing and measurement. Perdahcioğly et al. [3] found a linear relation between the amount of tension and the transformation rate in a 12Cr-9Ni-4Mo austenitic stainless steel. They concluded that the transformation rate increases proportionally with the increase in mechanical driving force, which agrees well with Patel and Cohen's theory on the action of applied stress on the martensitic transformation [26]. In this sense, in-situ measurements are a more suitable option to investigate the martensitic transformation kinetics. Both stress- and strain-induced types of martensitic transformation have been studied by means of complex, in-situ techniques such as neutron diffraction or HEXRD [8,9,11]. The information provided by these techniques is very accurate and reliable, but it is also localized. The inductive sensor used in this work is also able to measure both strain- and stress-induced types of transformation and can analyze sample volumes of 31 mm $\times$ 20 mm $\times$ 0.3 mm. Regarding the formation of stress induced martensite, insight can be obtained through LUR tests. In each unload/reload cycle, the austenite is stressed below its current flow stress. Therefore, if stress-induced transformation occurred during this stage, it would be reflected as an increment in the voltage after unloading/reloading. As Figure 6c displays, the voltage values before and after the unloading/reloading cycles do not match perfectly. A small increment can be appreciated and thus it could be attributed to the formation of stress-induced martensite during the cycle. However, its contribution to the whole transformation is very weak. For this reason, it can be said that the strain-induced mechanism prevails throughout the whole transformation in this case.

4.2. The Strain-Induced Martensitic Transformation and Stepwise Behavior

It is generally observed that the formation of ε-martensite during straining is favored at low temperatures and enhances the nucleation of α'-martensite [4,10]. The volume fraction of ε-martensite usually reaches a saturation of 0.02–0.05 at applied strains of 0.05–0.20, depending on the austenite grain size, and with larger strains the fraction decreases [7,9]. This suggests that it is a transient phase that represents a potential increase of nucleation sites for α' and, lowering the barrier for the nucleation of α'. Tian and co-workers have recently hypothesized that the transformation occurs via the sequence $\gamma \rightarrow$ stacking faults $\rightarrow \varepsilon \rightarrow \alpha'$ in MASSs with low SFE. Band-like structures resulting from faulted areas in the austenite act as precursors for ε-martensite. Although these observations correspond to the transformation during incremental cooling, the sequence could also be applied to the strain-induced transformation [11]. Other authors, on the other hand, conclude that ε-martensite and α'-martensite form independently during straining [14,16]. From the XRD results presented in Figure 2, it is observed that a small amount of ε-martensite forms along with α'-martensite during straining. Its presence is detected even in those conditions strained to 0.40–0.45, which could be interpreted as the ε-martensite not acting as precursor of α'-martensite in this case. Otherwise, a decrease in its volume fraction should be observed. This observation agrees well with Hedström and co-authors' work [16], who found by HEXRD during in-situ tensile testing in an AISI 301 grade that the formation of ε- and α'-martensite phases are not coupled. Instead, the volume fraction of ε-martensite reaches a saturation level and remains quite stable afterwards. Similarly, in the present work, both phases appear to form simultaneously with straining.

From the microstructural characterization presented in Figure 3, it is observed that the α'-martensite phase forms first at parallel slip bands constrained in the same austenite grain. Then, as the strain increases, more slip systems are activated and shear bands appear when homogeneous dislocation slip is inhibited or when an insufficient number of crystallographic slip systems is available in narrow deformation zones (order of microns), while the rest of the matrix undergoes comparably low and homogeneous plastic flow [18]. The number of intersections increases with further strain and both the nucleation and coalescence of α'-embryos along and across shear bands lead to a rise of the volume fraction of α'-martensite, as previously found [7]. This explains the overall transformation kinetics observed in Figure 9a, but it does not give answer to the step-like behavior pointed out in Figure 9b. Each step is formed due to a burst of α'-martensite within a very small strain increment. Hedström et al. [16] observed a correlation between sudden increases in martensite volume fraction and microstrains in austenite when studying an AISI 301 grade by HEXRD upon tensile loading. They attribute this effect to the autocatalytic martensitic transformation initiated by the growth and coalescence of martensite embryos.

Here, in order to explain the stepwise phenomenon, the relationship between microstructural development, martensite transformation kinetics, and work-hardening is discussed. The following analysis based on Hollomon's methodology [27] is carried out. Figure 10a presents the true stress (σ_T), and the instantaneous work-hardening rate ($\theta = d(\sigma_T)/d(\varepsilon_T)$) as a function of ε_T in the primary y-axis, and the fraction of strain-induced martensite ($f_{\alpha'}$) as a function of ε_T in the secondary y-axis. Figure 10b overlaps the evolution of θ with the martensite transformation rate ($df_{\alpha'}/d\varepsilon_T$) as a function of ε_T. It is observed that a volume fraction of about $f_{\alpha'} \approx 0.32$ has been formed when the stepwise behavior begins. Talonen et al. [28] explains this behavior in terms of the percolation theory. They found that a continuous network of α'-martensite forms above a $f_{\alpha'} \approx 0.32$, which was termed the percolation threshold. This theory is strongly supported by the microstructural evolution observations showed in Figure 3. It is easy to visualize that the formation of slip bands, martensite laths, and shear bands lead to a segmentation of the untransformed austenite phase into smaller and smaller volumes. At the same time, the flow stress of the austenite increases gradually due to work-hardening. The combination of these two effects: (1) austenite volume refinement and (2) flow stress increase, increases the elastic energy needed for the deformation of the matrix and the surface energy required to create a new interface. In turn, the mechanical driving force for the transformation increases. It can be thus deduced that the resulting microstructure consists of a continuous network of α'-martensite with dispersed untransformed islands of austenite, similar to the configuration first described by Talonen [28]. As Figure 10b depicts, this threshold approximately corresponds to the highest transformation rate, and afterwards, the rate continuously decreases while experiencing pulses of very high intensity. Once the percolation threshold is reached, the load cannot be fully accommodated by the austenite and it is transferred to the α'-martensite. Due to a non-homogenous plastic deformation, the accumulation of stresses is enhanced within the untransformed austenite islands, which triggers the autocatalytic (or stress-induced) martensitic transformation in this volume [1,3]. The occurrence of these bursts in a gradual manner might be related to the crystallographic orientation of the austenite grains with respect to the applied stress, which influences the stress at which the transformation starts. After each burst of sudden martensite formation there is a plateau where almost no transformation occurs. This might be due to the exothermic nature of the martensitic transformation. After each burst, the local temperature of the sample increases, which has two effects: (1) the decrease of the chemical driving force for the transformation ($\Delta G^{\gamma \rightarrow \alpha'}$) [11], and (2) the increase of the SFE [4]. Both effects hinder the transformation, which might explain the plateaus. The variation of the transformation rate in pulses causes localized hardening, which redistributes the load to another part of the sample. As a result, the occurrence of necking is pushed to higher strain levels and a high hardening effect is provided. This might be the reason why the uniform and total elongations are found at very similar values.

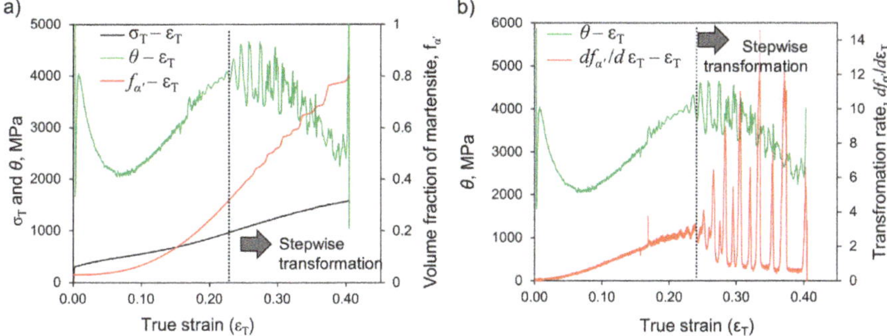

Figure 10. (**a**) Evolution of true stress (σ_T), instantaneous work-hardening rate (θ) and $f_{\alpha'}$ with true strain (ε_T). (**b**) Evolution of θ and martensite formation rate ($df_{\alpha'}/d\varepsilon_T$) with ε_T. The dotted line highlights the origin of the stepwise transformation.

4.3. Validation of the Inductive Sensor as a Technique for In-Situ Characterization of the Strain-Induced Martensitic Transformation

Contrary to other techniques, the sensor is an easy-to-use tool, which does not consume extra measurement-time, need extra sample preparation, or require large facilities. The correlation between the inductive signal and the α'-martensite volume fraction has allowed for the validation of the sensor developed by Philips ATC [29] as an in-situ characterization technique to study the progress of the strain-induced martensitic transformation in MASSs. It is known that the calibration of magnetic devices to quantify volume fractions of martensite can be very complex [6,15,30]. The influence of some effects on the α'-martensite permeability makes the post-processing of the output signal necessary in order to make it representative of the volume fraction of martensite present in the material. In this work, both the effect of temperature and applied stress were minimized by lowering the current circulating through the coil to 55 mA, as shown in Figure 6a. A current of 55 mA ensures a magnetic field of 0.6 T, which is sufficiently high to generate a magnetization response representative of the martensite volume fraction in the material but sufficiently low to not influence the martensitic transformation during the time that the monotonic tensile tests take, which is approximately 9 min [23]. Besides, this current does not warm up the sample. Although the stress effect is significantly reduced, it is still present in the signal and becomes larger and larger as the volume fraction of α'-martensite increases above 0.45 ($\varepsilon_E{}^{th} > 0.31$). The same effect is observed for larger current values (Figure 6a,b). Therefore, the correction of the stress effect is needed regardless the current circulating through the coil; however, as it has been mentioned previously, an optimum value of 55 mA does not heat up the sample and thus does not influence the driving force for the martensitic transformation.

5. Conclusions

The inductive sensor developed by Philips ATC has been proven to be a user-friendly, cheap, and fast tool for characterizing the formation of α'-martensite in metastable austenitic stainless steels during tensile testing. The calibration of the sensor has been facilitated by setting the current circulating through the coil to 55 mA, which ensures a magnetic field of 0.6 T. This magnetic field is sufficiently high to magnetize the sample, but avoids any influence on the transformation kinetics, either by heating-up of the sample or by accelerating the formation of martensite.

The formation of strain-induced α'-martensite has been studied at room temperature in an AISI 301 grade. In-situ inductive measurements performed with the sensor were compared to the ex-situ characterization done on interrupted-tensile tested specimens by means of light optical microscopy, X-ray diffraction and magnetization measurements. The metallographic quantification of the α'-martensite volume fraction leads to an overestimation; whereas XRD and SQUID magnetization measurements are in good agreement. It seems that SQUID magnetization measurements lead to a more sensitive detection of the α'-martensite volume fraction and match well with the response obtained with the inductive sensor. A correlation between the sensor output signal and the volume fraction of α'-martensite was found, which will be used in future experiments.

The in-situ study has revealed that the α'-martensite transformation occurs in steps once the percolation threshold is reached. At this point the microstructure consists of a continuous network of α'-martensite and islands of untransformed austenite. The phenomena occurring during the martensitic transformation during deformation, i.e., formation of slip bands, shear bands, and martensite laths, produces the segmentation of austenite grains in smaller and smaller volumes. This effect, along with the increase in austenite flow stress due to work-hardening, results in an increase of the mechanical driving force for the transforming. As a consequence, the stresses accumulate in the austenite, which transforms rapidly as a consequence of the autocatalytic nature of the transformation. This rapid growth involves the coalescence of α'-martensite blocks in the continuous network, which causes bursts of martensitic transformation.

Acknowledgments: Carola Celada-Casero acknowledges the financial support from Consejo Superior de Investigaciones Científicas (CSIC) in the form of a JAE-Predoc grant (JAEPre_2011_01167), co-funded by the European Social Fund. The authors deeply acknowledge the financial support from Research Fund for Coal for funding this research under the Contract RFCS-02-2015 (Grant Agreement No.: RFSR-CT-2012-00021) and the Spanish Ministerio de Economia y Competitividad (MINECO) through the form of a Coordinate Project (MAT2016-80875-C3-1-R). Authors are also grateful to Jose Antonio Jimenez (CENIM-CSIC), Julio Romero de Paz (CAI of Physical Techniques, UCM) and Marijke de Vries and Ronald van der Linden (Philips ATC) for the experimental support.

Author Contributions: Carola Celada-Casero performed the experiments, analyzed the results and wrote the paper; Harm Kooiker and Manso Groen provided the material, the facilities and the tools for the analysis and data interpretation; Jan Post conceived and designed the experiments for the sensor calibration; David San-Martin interpreted the data, discussed the results and reviewed the paper.

Conflicts of Interest: The authors declare no conflict of interest.

References

1. Olson, G.B.; Cohen, M. A mechanism for the strain-induced nucleation of martensitic transformations. *J. Less Common Met.* **1972**, *28*, 107–118. [CrossRef]
2. Celada-Casero, C.; Huang, B.M.; Aranda, M.M.; Yang, J.R.; San Martin, D. Mechanisms of ultrafine-grained austenite formation under different isochronal conditions in a cold-rolled metastable stainless steel. *Mater. Charact.* **2016**, *118*, 129–141. [CrossRef]
3. Perdahcıoğlu, E.S.; Geijselaers, H.J.M.; Groen, M. Influence of plastic strain on deformation-induced martensitic transformations. *Scr. Mater.* **2008**, *58*, 947–950. [CrossRef]
4. Talonen, J.; Hänninen, H. Formation of shear bands and strain-induced martensite during plastic deformation of metastable austenitic stainless steels. *Acta Mater.* **2007**, *55*, 6108–6118. [CrossRef]
5. Celada-Casero, C.; Chao, J.; Urones-Garrote, E.; San Martin, D. Continuous hardening during isothermal aging at 723 k (450 °C) of a precipitation hardening stainless steel. *Metall. Mater. Trans. A* **2016**, *47*, 5280–5287. [CrossRef]
6. Perdahcıoğlu, E.S.; Geijselaers, H.J.M.; Huétink, J. Influence of stress state and strain path on deformation induced martensitic transformations. *Mater. Sci. Eng. A* **2008**, *481–482*, 727–731.
7. Kisko, A.; Misra, R.D.K.; Talonen, J.; Karjalainen, L.P. The influence of grain size on the strain-induced martensite formation in tensile straining of an austenitic 15cr-9mn-ni-cu stainless steel. *Mater. Sci. Eng. A* **2013**, *578*, 408–416. [CrossRef]
8. Haušild, P.; Davydov, V.; Drahokoupil, J.; Landa, M.; Pilvin, P. Characterization of strain-induced martensitic transformation in a metastable austenitic stainless steel. *Mater. Des.* **2010**, *31*, 1821–1827. [CrossRef]
9. Hedström, P.; Lindgren, L.E.; Almer, J.; Lienert, U.; Bernier, J.; Terner, M.; Odén, M. Load partitioning and strain-induced martensite formation during tensile loading of a metastable austenitic stainless steel. *Metall. Mater. Trans. A* **2009**, *40*, 1039–1048. [CrossRef]

10. Spencer, K.; Embury, J.D.; Conlon, K.T.; Véron, M.; Bréchet, Y. Strengthening via the formation of strain-induced martensite in stainless steels. *Mater. Sci. Eng. A* **2004**, *387–389*, 873–881.

11. Tian, Y.; Lienert, U.; Borgenstam, A.; Fischer, T.; Hedström, P. Martensite formation during incremental cooling of fe-cr-ni alloys: An in-situ bulk X-ray study of the grain-averaged and single-grain behavior. *Scr. Mater.* **2017**, *136*, 124–127. [CrossRef]

12. Murr, L.E.; Staudhammer, K.P.; Hecker, S.S. Effects of strain state and strain rate on deformation-induced transformation in 304 stainless steel: Part II Microstructural study. *Metall. Trans. A* **1982**, *13*, 627–635. [CrossRef]

13. Talonen, J.; Hänninen, H.; Nenonen, P.; Pape, G. Effect of strain rate on the strain-induced $\gamma \rightarrow \alpha'$-martensite transformation and mechanical properties of austenitic stainless steels. *Metall. Mater. Trans. A* **2005**, *36*, 421–432. [CrossRef]

14. Lichtenfeld, J.A.; Van Tyne, C.J.; Mataya, M.C. Effect of strain rate on stress-strain behavior of alloy 309 and 304l austenitic stainless steel. *Metall. Mater. Trans. A* **2006**, *37*, 147–161. [CrossRef]

15. Post, J.; Nolles, H.; Datta, K.; Geijselaers, H.J.M. Experimental determination of the constitutive behaviour of a metastable austenitic stainless steel. *Mater. Sci. Eng. A* **2008**, *498*, 179–190. [CrossRef]

16. Hedström, P.; Lienert, U.; Almer, J.; Odén, M. Stepwise transformation behavior of the strain-induced martensitic transformation in a metastable stainless steel. *Scr. Mater.* **2007**, *56*, 213–216. [CrossRef]

17. Celada Casero, C.; San Martín, D. Austenite formation in a cold-rolled semi-austenitic stainless steel. *Metall. Mater. Trans. A* **2014**, *45*, 1767–1777. [CrossRef]

18. Jia, N.; Eisenlohr, P.; Roters, F.; Raabe, D.; Zhao, X. Orientation dependence of shear banding in face-centered-cubic single crystals. *Acta Mater.* **2012**, *60*, 3415–3434. [CrossRef]

19. Olson, G.B.; Cohen, M. Kinetics of strain-induced martensitic nucleation. *Metall. Trans. A* **1975**, *6*, 791–795. [CrossRef]

20. Shin, H.C.; Ha, T.K.; Chang, Y.W. Kinetics of deformation induced martensitic transformation in a 304 stainless steel. *Scr. Mater.* **2001**, *45*, 823–829. [CrossRef]

21. Post, J.; Huetink, J.; Geijselaers, H.J.M.; Voncken, R.M.J. Fem simulations of a multi stage forming process on sandvik maraging steel 1rk91 describing the stress assisted and the strain induced martensite formation. *J. Phys. IV* **2003**, *112*, 417–420. [CrossRef]

22. Zhao, L.; van Dijk, N.H.; Brück, E.; Sietsma, J.; van der Zwaag, S. Magnetic and X-ray diffraction measurements for the determination of retained austenite in trip steels. *Mater. Sci. Eng. A* **2001**, *313*, 145–152. [CrossRef]

23. San Martín, D.; Aarts, K.W.P.; Rivera-Díaz-del-Castillo, P.E.J.; van Dijk, N.H.; Brück, E.; van der Zwaag, S. Isothermal martensitic transformation in a 12cr-9ni-4mo-2cu stainless steel in applied magnetic fields. *J. Mag. Mag. Mater.* **2008**, *320*, 1722–1728. [CrossRef]

24. Mumtaz, K.; Takahashi, S.; Echigoya, J.; Kamada, Y.; Zhang, L.F.; Kikuchi, H.; Ara, K.; Sato, M. Magnetic measurements of martensitic transformation in austenitic stainless steel after room temperature rolling. *J. Mater. Sci.* **2004**, *39*, 85–97. [CrossRef]

25. De Abreu, H.F.G.; da Silva, M.J.G.; Herculano, L.F.G.; Bhadeshia, H. Texture analysis of deformation induced martensite in an aisi 301l stainless steel: Microtexture and macrotexture aspects. *Mater. Res.* **2009**, *12*, 291–297. [CrossRef]

26. Patel, J.R.; Cohen, M. Criterion for the action of applied stress in the martensitic transformation. *Acta Metall.* **1953**, *1*, 531–538. [CrossRef]

27. Hollomon, J.H. Tensile deformation. *Trans. Metall. Soc. AIME* **1945**, *162*, 268–290.

28. Talonen, J. *Effect of Strain-Induced α'-Martensite Transformation on Mechanical Properties of Metastable Austenitic Stainless Steels*; Helsinki University of Technology: Helsinki, Finland, 2007.

29. Post, J. *On the Constitutive Behaviour of Sandvik Nanoflex TM—Modelling Experiments and Multi-Stage Forming*; University of Twente: Enschede, The Netherlands, 2004.

30. Radu, M.; Valy, J.; Gourgues, A.F.; Strat, F.L.; Pineau, A. Continuous magnetic method for quantitative monitoring of martensitic transformation in steels containing metastable austenite. *Scr. Mater.* **2005**, *52*, 525–530. [CrossRef]

Article

Phase Equilibrium and Austenite Decomposition in Advanced High-Strength Medium-Mn Bainitic Steels

Adam Grajcar [1],*, Władysław Zalecki [2], Wojciech Burian [2] and Aleksandra Kozłowska [1]

[1]   Faculty of Mechanical Engineering, Silesian University of Technology, 18a Konarskiego Street, Gliwice 44-100, Poland; aleksandra.kozlowska@polsl.pl

[2]   Institute for Ferrous Metallurgy, 12-14 K. Miarki Street, Gliwice 44-100, Poland; wzalecki@imz.pl (W.Z.); wburian@imz.pl (W.B.)

*   Correspondence: adam.grajcar@polsl.pl; Tel.: +48-32-237-2940

Academic Editor: Carlos Garcia-Mateo
Received: 15 September 2016; Accepted: 14 October 2016; Published: 20 October 2016

Abstract: The work addresses the phase equilibrium analysis and austenite decomposition of two Nb-microalloyed medium-Mn steels containing 3% and 5% Mn. The pseudobinary Fe-C diagrams of the steels were calculated using Thermo-Calc. Thermodynamic calculations of the volume fraction evolution of microstructural constituents vs. temperature were carried out. The study comprised the determination of the time temperature-transformation (TTT) diagrams and continuous cooling transformation (CCT) diagrams of the investigated steels. The diagrams were used to determine continuous and isothermal cooling paths suitable for production of bainite-based steels. It was found that the various Mn content strongly influences the hardenability of the steels and hence the austenite decomposition during cooling. The knowledge of CCT diagrams and the analysis of experimental dilatometric curves enabled to produce bainite-austenite mixtures in the thermomechanical simulator. Light microscopy (LM), scanning electron microscopy (SEM), and transmission electron microscopy (TEM) were used to assess the effect of heat treatment on morphological details of produced multiphase microstructures.

Keywords: medium-Mn steel; austenite decomposition; dilatometry; phase equilibrium; retained austenite

1. Introduction

The beneficial combination of high strength, ductility, and formability of steel sheets for the automotive industry can be achieved using advanced high-strength steels (AHSS). They consist of different soft and hard structural constituents in various proportions, which enable the attainment of a very wide range of mechanical and technological properties. The microstructure of dual-phase (DP) steel consists of soft ferrite and hard martensite, whereas the multiphase microstructure of TRIP (transformation-induced plasticity) steel comprises ferrite, bainite, and retained austenite [1–6]. New demands of the automotive industry for relatively low-cost steel sheets characterized by tensile strength above 1000 MPa require further searching for new chemical composition strategies. Advanced ultrahigh strength steels contain a higher fraction of hard phases (i.e., acicular ferrite, bainite, or martensite [7–10]) compared to AHSS containing polygonal ferrite as a matrix.

A key microstructural constituent of advanced multiphase steels is retained austenite in amounts from 10% to 30%. The latter phase ensures a required ductility level by its strain-induced martensitic transformation during cold-forming operations. Recently, a high amount of retained austenite can be obtained in different bainitic alloys containing from 1.5 to 8 wt. % of Mn, which is a main austenite stabilizer [8–14]. These steels are dedicated to the automotive industry for different crash-relevant elements, especially in the side zone of a car (B-pillars, roof rails, side-impact beams, etc.). However, their wide use requires improvement of forming technologies and special welding procedures.

Monitoring the phase transformations and the knowledge of continuous cooling transformation (CCT) diagrams are of primary importance for proper design of bainite-austenite microstructures in medium-Mn steels. Austenite decomposition upon heating or during cooling from the γ region is often monitored by dilatometry, differential thermal analysis (DTA), or differential scanning calorimetry (DSC) [11,15–17]. Results of these investigations have to be confirmed by detailed microscopic research because phase transformations in multiphase steels are very complex. In medium-Mn steels, the bainite is particularly difficult for unequivocal identification because it can contain carbides or may form carbide-free bainite [8,18–22]. Films of retained austenite, instead of carbides, occur between laths of bainitic ferrite.

The destabilization of the austenite occurs during heating or cooling as a result of precipitation of carbides, martensitic transformation, and so forth. Monitoring the volume fraction of all microstructural constituents and their morphology is key to obtaining optimal mechanical properties of multiphase steels. Beneficial mechanical properties and formability of steels with a bainitic-austenitic mixture are obtained for fine, homogeneous bainite microstructures. Carbide precipitates and a bimodal morphology of fine and coarse bainite are detrimental for fracture toughness and ductility of steel products [19,20,23]. Thus, silicon or aluminum strategies are employed to avoid carbide precipitation during isothermal bainitic transformation step [24–27].

2. Materials and Methods

2.1. Materials

Medium-Mn steels contain from 3% to 12% of Mn. Steel with higher Mn contents are used mainly for cold-rolled products [3,10,12], but the steels relevant to the current study were being designed for hot-rolled products. Hence, the steels contained the lower levels of Mn (i.e., in a range from 3 to 5 wt. %). The carbon content in both steels was at a level of 0.17 wt. % (Table 1), which provides moderate hardenability. Aluminium concept was chosen to prevent carbide precipitation. Mo and Nb were added for solid solution hardening and precipitation strengthening. The alloys differed only in the manganese content, thus they were coded as 3 Mn and 5 Mn steels. The laboratory ingots, after vacuum induction melting, were hot-forged to obtain flat samples with a thickness of 22 mm. Cylindrical (solid) samples of $\phi 5 \times 7$ mm for dilatometric tests and cuboid samples of $15 \times 20 \times 35$ mm^3 for multistep compression tests were machined.

Table 1. Chemical compositions of the investigated medium-Mn steels.

Species	Grade								
	C %	Mn %	Al %	Si %	Mo %	Nb %	P %	S %	N %
3 Mn	0.17	3.1	1.6	0.22	0.22	0.04	0.008	0.005	0.0046
5 Mn	0.17	5.0	1.5	0.21	0.20	0.03	0.008	0.005	0.0054

2.2. Thermodynamic Calculations

The investigated steels belong to a new group of unconventional iron alloys due to untypical additions of Mn and Al. Hence, the precise selection of heat treatment or thermomechanical-processing conditions requires earlier thermodynamic calculations of phase compositions and other parameters. The first step required determination of pseudobinary iron-carbon systems for the investigated alloys using the Thermo-Calc package. The calculations of phase diagrams were performed by means of Thermo-Calc software using TCFE6 and SSOL2 databases [28]. The untypical alloying additions can substantially affect the decomposition of austenite upon cooling [29]. Thus, the time-temperature-transformation (TTT) diagrams for isothermal cooling conditions and continuous cooling transformation (CCT) diagrams were calculated using the JMatPro (Sente Software Ltd., Guildford, UK) software [30]. This package was also used to monitor the evolution of potential phases as a function of temperature.

2.3. Experimental

The correctness of thermodynamic calculations was checked by dilatometric analysis and microstructure studies. Critical temperatures were determined using tubular samples of $\phi 4 \times \phi 2 \times 7$ mm, by means of a DIL805A/D dilatometer manufactured by Bähr Thermoanalyse GmbH (Hüllhorst, Germany). The solid samples of $\phi 5 \times 7$ mm in the dilatometric tests with deformation were heated under vacuum conditions (vacuum $< 10^{-4}$ mbar) to 1100 °C at a rate of 3 °C/s. After 300 s soaking at this temperature they were cooled to 900 °C at a rate of 5 °C/s. The holding time before deformation and subsequent cooling was 20 s. Then the samples were compressed, applying the true strain of 0.5 at a strain rate of 1/s. The continuous cooling rates to room temperature were selected as 0.5, 1, and 4 °C/s.

The final stage of the experimental study comprised the simulation of hot-working and cooling conditions used for production of thermomechanically processed TRIP steels with retained austenite. These conditions were reconstructed in the Gleeble 3800 thermomechanical simulator (Dynamic Systems Inc., Poestenkill, NY, USA). The initial processing schedule included resistance-heating of cubicoid samples to 1200 °C and their soaking within 30 s. Then, the samples were deformed using seven compression steps, summarized in Table 2. After the final deformation at 850 °C, the specimens were controlled cooled to room temperature to produce different multiphase microstructures. Since the decomposition of austenite substantially affects CCT and TTT diagrams, the cooling conditions will be presented in the next section after analysing phase transformations of undercooled austenite.

Table 2. Hot-working conditions.

Compression Step	Deformation Temperature, °C	True Strain	Strain Rate, s^{-1}	Time between Compression Steps, s
1	1150	0.4	5	30
2	1000	0.3	5	10
3	950	0.2	10	10
4	900	0.2	10	3
5	870	0.2	30	3
6	860	0.2	50	1
7	850	0.2	50	–

Light microscopy (LM), scanning electron microscopy (SEM), and transmission electron microscopy (TEM) were employed to identify phase composition of the investigated steels. Standard techniques for the preparation of metallographic samples were applied. Nital etching was used for dilatometric specimens whereas etching in 10% aqueous solution of sodium metabisulphite was applied to reveal retained austenite in thermomechanically processed samples. Morphological details of the microstructure were revealed with the SUPRA 25 SEM (Carl Zeiss, Jena, Germany). The quantitative metallography has been applied to determine the amount of ferrite fraction in dilatometric specimens. The presence of potential precipitations and substructure morphology were identified using JEOL JEM 2000 FX (JEOL USA, Inc., Peabody, MA, USA) electron microscope at an accelerating voltage of 160 kV.

3. Results and Discussion

3.1. Thermodynamic Calculations

The pseudobinary Fe-C phase diagram of 3 Mn steel determined using the Thermo-Calc is shown in Figure 1. It is clearly seen that the shape of the diagram in the high-temperature region is far from typical low-alloyed steels. The solidification of the alloy begins with crystallization of δ ferrite from the liquid. The high addition of Al increased a maximum solubility of C in the δ ferrite to ca. 0.17 wt. %, corresponding to its content in the alloy. It is also obvious that for steels containing below 0.3 wt. % C, the stability of pure austenite cannot be obtained after passing through the two-phase region. Instead, the stability of austenite and ferrite occurs up to a eutectoid reaction at about 700 °C. This phenomenon is caused by the addition high amounts of Al, which is a strong ferrite former. Precipitation of niobium carbides starts to take place at a temperature of about 1200 °C under conditions of thermodynamic equilibrium.

The addition of 5 wt. % Mn slightly changes the pseudobinary Fe-C phase diagram, which is presented in Figure 2. As a result of the higher Mn addition, the solubility of C in δ ferrite decreased to approx. 0.12 wt. %. This means that some fraction of γ phase should be formed from the liquid. Manganese also shifts the two-phase region of γ + α to the left, but pure austenite still cannot be produced due to the opposite effect of Al. A temperature of equilibrium cementite precipitation decreased below 700 °C.

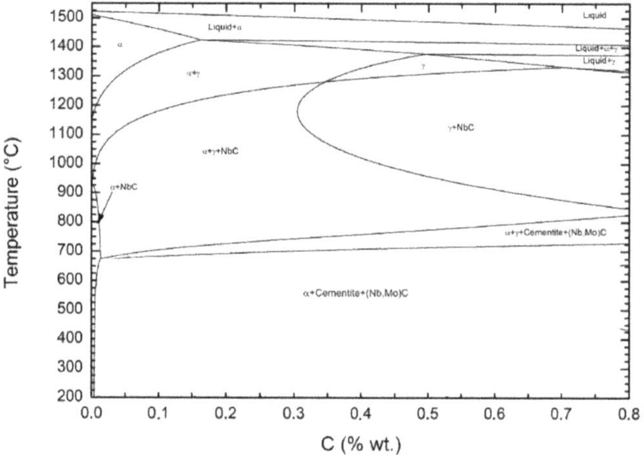

Figure 1. Pseudobinary Fe-C phase diagram of 3 Mn steel determined using Thermo-Calc.

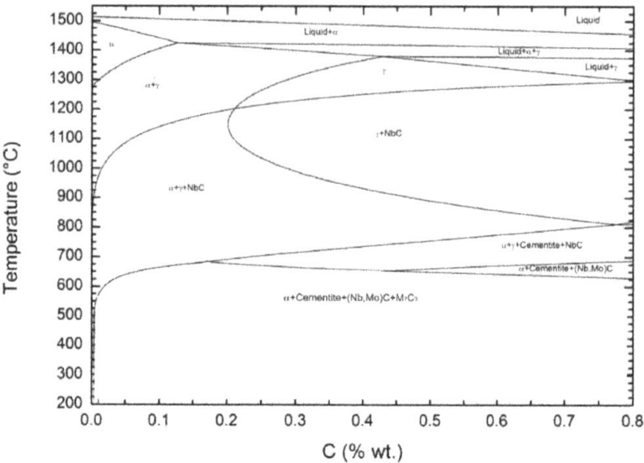

Figure 2. Pseudobinary Fe-C phase diagram of 5 Mn steel determined using Thermo-Calc.

Further details of the microstructure evolution as a function of temperature under equilibrium conditions can be assessed on the basis of Figures 3 and 4. Figure 3 shows the evolution of the potential structural constituents in the steel containing 3 wt. % of Mn. One can see that the solidification process is accompanied by precipitation of manganese sulphides. However, their amount should be relatively small, taking into account the high metallurgical cleanliness of the steels, as sulphur contents do not exceed the limit of ca. 0.005% (Table 1).

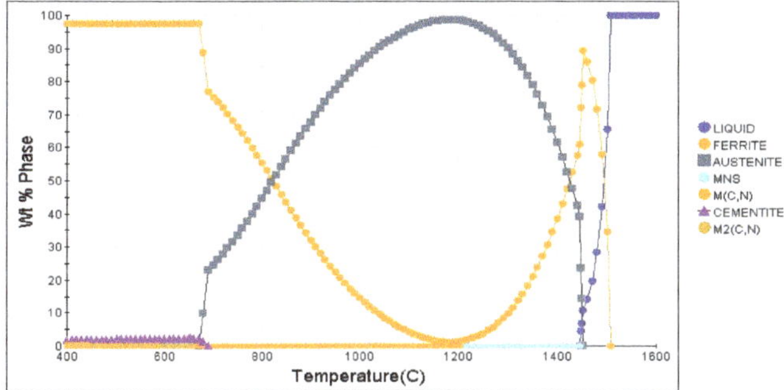

Figure 3. Equilibrium evolution of phases in 3 Mn steel calculated using the JMatPro package.

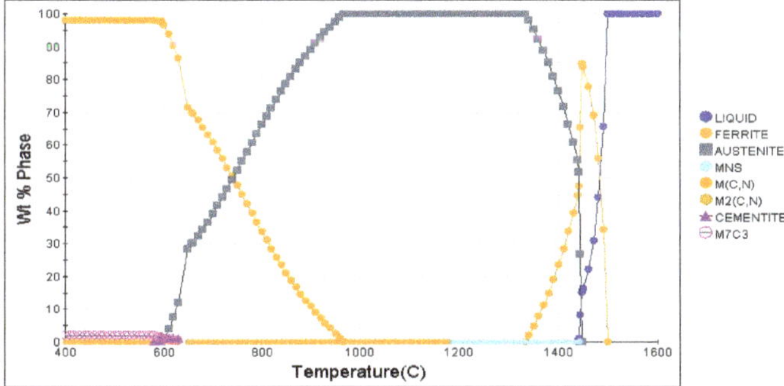

Figure 4. Equilibrium evolution of phases in 5 Mn steel calculated using the JMatPro package.

Figure 3 confirms the prediction of the pseudobinary Fe-C system of the 3 Mn steel concerning a lack of stable single γ phase. The maximum content of austenite with a small fraction of ferrite occurs under the equilibrium at a temperature of ca. 1180 °C. The α phase amount increases with the temperature drop. Below the eutectoid reaction cementite will be formed. The potential presence of M(C,N) and/or M_2(C,N) precipitates cannot be excluded.

The higher discrepancy between the Fe-C system (Figure 2) and the evolution of predicted phases occurs for the 5 Mn steel. Figure 4 suggests that there is a stable austenite field in a temperature range from ca. 1320–960 °C. On the other hand, Figure 2 does not contain the stable austenite region. However, it should be noted that for the C content of 0.17 wt. % this line is situated close to the transition line (γ + NbC). It seems that for new chemical compositions further improvements in Thermo-Calc and JMatPro packages are needed to predict more correctly the phase transitions of such complex systems. This especially concerns the combined effects caused by increased Mn and Al contents. These predictions have been validated partially in dilatometric tests performed, including plastic deformation of austenite at 900 °C prior to cooling. The results of these tests will be presented in Section 3.2. Further experiments are planned for long holding times at higher austenitizing temperatures followed by sample quenching to correlate the real phase state with equilibrium predictions.

Industrial technology development requires investigation of the microstructure evolution as a function of time. Thus, Figures 5 and 6 present predicted time-temperature-transformation (TTT) diagrams for both steels. The A_{e3} and A_{e1} temperatures are equal to 964 and 694 °C, respectively, for the 3 Mn steel, and they are substantially smaller for the steel containing 5 wt. % Mn: $A_{e3} = 874$ °C and $A_{e1} = 655$ °C. The evident effect of the high-Mn contents on the displacement of all phase transformations to a longer time is visible. This effect is especially significant for diffusion-controlled transformations. For example, the ferrite transformation begins after 100 s at 800 °C (Figure 5). In the case of the 5 Mn steel, the delay for the same isothermal holding temperature at which the ferrite is formed is even 10 times longer (Figure 6). It appears that the substantial hardenability effect of Mn suppresses the production of ferrite-based structures after cooling from an austenitization temperature.

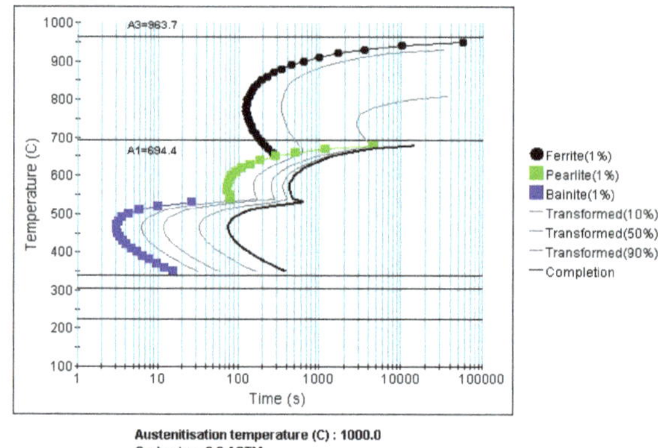

Figure 5. Predicted time-temperature-transformation (TTT) diagram of 3 Mn steel determined using the JMatPro package.

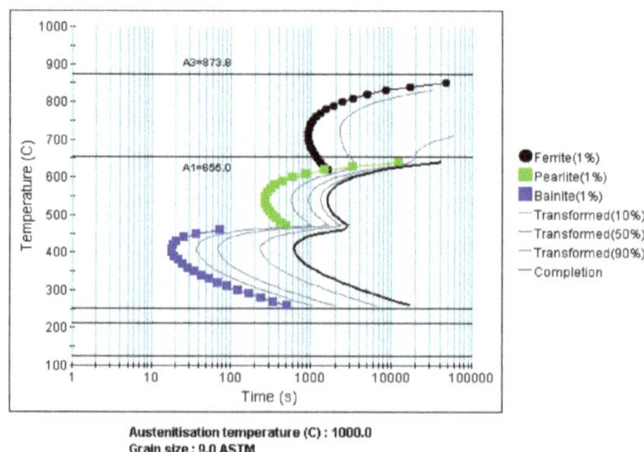

Figure 6. Predicted TTT diagram of 5 Mn steel determined using the JMatPro package.

It should be noted that both for isothermal and continuous cooling conditions, TTT and CCT diagrams are dependent on prior austenite grain size. For the present calculations the initial grain size of 9 ASTM (American Society for Testing and Materials, ASTM International) has been taken, which corresponds to a size of ca. 14 μm. An austenitizing temperature of 1000 °C has been assumed. The selected temperature and grain size may seem to be too low. However, the aim was to take into account a grain size after plastic deformation, which is usually successfully decreased during multistep deformation tests [21,24].

The pearlite transformation starts after ca. 70 s and 250 s at 560 °C, respectively, for the 3 Mn and 5 Mn steels. This delay is beneficial in terms of the production of carbide-free microstructures. The martensite start temperatures are relatively low: 340 °C for the 3 Mn steel and 251 °C for the alloy with the higher Mn content. The most interesting are bainitic bays in both steels because they constitute large technological temperature-time windows for realization of isothermal bainitic treatments, which are needed for production of carbide-free bainite-austenite mixtures [7,8,27].

A very similar situation is observed in the case of the continuous cooling transformation (CCT) diagrams (Figures 7 and 8). All regions of supercooled austenite transformations are shifted to the right. The displacement of ferrite allows supposition that it will be difficult to induce some fraction of this phase in hot-rolled sheets. This assumption has been confirmed in our earlier initial tests involving real dilatometric tests, which enabled the construction of experimental CCT diagrams [31]. The calculated and real CCT diagrams are in good agreement. However, further work is intended to explain in detail the presence of different phases, the effect of strain, composition, and cooling rates.

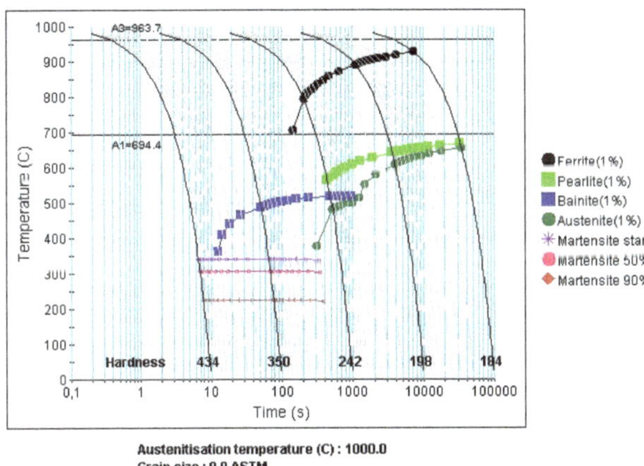

Figure 7. Predicted continuous cooling transformation (CCT) diagram of 3 Mn steel determined using the JMatPro package.

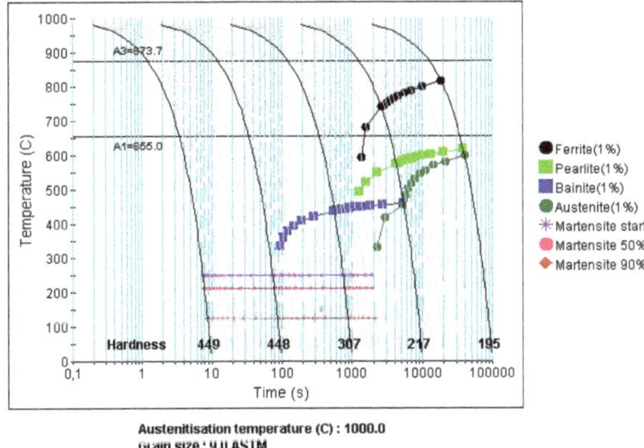

Figure 8. Predicted CCT diagram of 5 Mn steel determined using the JMatPro package.

It should be taken into account that plastic deformation shifts diffusion-controlled transformations to a shorter time [31,32]. JMatPro software does not enable the user to take into account the influence of inner energy increase due to plastic deformation on phase transformations. Only a change of grain size may be applied. However, it requires some extra investigation of rheology of the material to gain reliable data. Thus, this behaviour requires experimental verification both in terms of ferrite production and carbide-free bainite formation.

The formation of bainite from the strained austenite is more complex compared to the ferrite transformation. The reason for that is a critical temperature region of bainitic transformation, at which diffusive and diffusionless transitions can take place. Moreover, the transformation behaviour strongly depends on a stress state. Bhadeshia [33] and Hase et al. [34] showed that, on the one hand, compressive or tensile stresses increase the bainite start temperature and the resulting transformation rate. On the other hand, the hydrostatic compression decreases the bainite formation rate. Considering the similarity of the bainitic transformation and martensite formation (in a low-temperature bainitic region) one can expect that deformation debris in the austenite retards the growth of bainite, causing a reduction in the fraction of transformation in spite of an increased number density of nucleation sites. This is called mechanical stabilization of bainite. At the same time, this phase is more refined [35]. Hence, the transformation rate is at first promoted by strain, and then it is retarded. This effect is stronger in a low-temperature range of bainitic transformation. Moreover, it should be noted that the austenite is usually inhomogeneously deformed. As a result, the lightly deformed regions transform rapidly because of the increase in the defect density, whereas the overall transformed area of the strongly deformed regions is smaller [33]. Finally, it results in bimodal distribution of bainite subunit sizes. The refinement of blocky austenite grains and the inherited increased dislocation density can support the stabilization of retained austenite within the bainitic matrix [12,35].

3.2. Dilatometric Tests

The representative dilatometric curves of the steel samples plastically deformed at 900 °C and then continuously cooled to room temperature at a rate of 1 °C/s are shown in Figures 9 and 10. Figure 9 presents dilatation changes for the steel containing the lower Mn content. Determination of ferrite start temperature is not easy because the deviation on the dilatometric curve is very small. Thus, developed in the Institute for Ferrous Metallurgy, a method of linear transformation of dilatometric curves was applied to get rid of "scale effect". The method of curve differentiation was also used, but the results were not good enough because of the scattering of experimental data. For the presented exemplary dilatometric curve in Figure 9, the phase composition comprised ca. 67% of bainite, 18% of martensite, and 15% of ferrite after applying a lever method. The A_{r3} temperature was determined to be 789 °C.

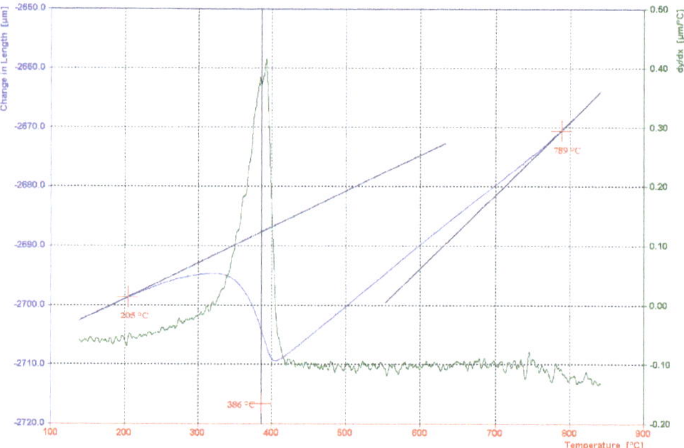

Figure 9. Dilatometric curve registered during cooling of 3 Mn steel at a rate of 1 °C/s.

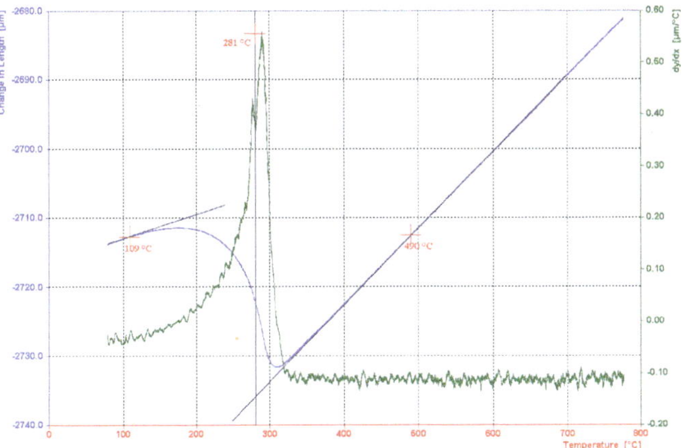

Figure 10. Dilatometric curve registered during cooling of 5 Mn steel at a rate of 1 °C/s.

The amount of austenite transformed into ferrite can be seen in Figure 11b. The average amount of ferrite—determined using an image analysis—contains 8.7% ± 1.6%, 8.5% ± 1.7%, and 2.3% ± 1.1%, respectively for the cooling rates of 0.5 °C/s, 1 °C/s, and 4 °C/s. Moreover, it is difficult to distinguish between a product of the austenite transformation and primary ferrite. The occurrence of the latter one confirms the correctness of Thermo-Calc predictions in Figure 1. Moreover, the fraction of large primary ferrite grains does not exceed ca. 7%, which is in good agreement with Figure 3. The application of the slower cooling rate amounting to 0.5 °C/s does not result in the essential increase of a fraction of ferrite (Figure 11a). This phase is still embedded in the bainite-martensite matrix. For faster cooling rates, only primary ferrite exists together with bainitic-martensitic mixture (Figure 11c).

The analysis of the dilatometric curve of the 5 Mn steel is easier because of a lack of an expansion signal from ferrite (Figure 10). The detailed analysis allowed estimating the bainite start temperature as ca. 490 °C. The martensite start temperature determined using differentiation of the curve was equal to 269 °C. These temperatures are significantly lower compared to the 3 Mn steel, which should be ascribed to the hardenability effect of manganese. The corresponding micrographs for the cooling rates from 0.5 °C/s to 4 °C/s are presented in Figure 11d–f. It is clear that these are ferrite-free bainitic-martensitic microstructures with the amount of martensite, which increases with faster cooling rates. A lack of primary ferrite corresponds well with the JMatPro calculations in Figure 4 and indicates some discrepancy between the experiment and calculations concerning the pseudobinary Fe-C system (Figure 2).

3.3. Microstructure of Thermomechanically Processed Samples

The phase transition calculations and dilatometric tests were the basis for the construction of final cooling schedules after multistep compression tests (Table 2). The results of analysis of dilatometric tests have made it clear that the present medium-Mn steels are not prone to obtaining ferrite-based microstructures. Therefore, the main cooling paths were designed in terms of the production of bainite-based microstructures. It should be noted that the applied cooling rates and deformation patters were different when compared to the dilatometric tests, which was obvious in terms of the design of multistep hot-strip rolling schedules. However, some fundamental indications on the effects of plastic deformation and cooling conditions on the transformation behaviour have been known. The most important information gained is regarding the difficulties in production of polygonal ferrite. The detailed cooling schedules are listed in Table 3. The essential step for stabilization of retained austenite is isothermal holding of deformed specimens at 450 °C within 300 s, where an incomplete bainitic transformation phenomenon should take place [16,23,27].

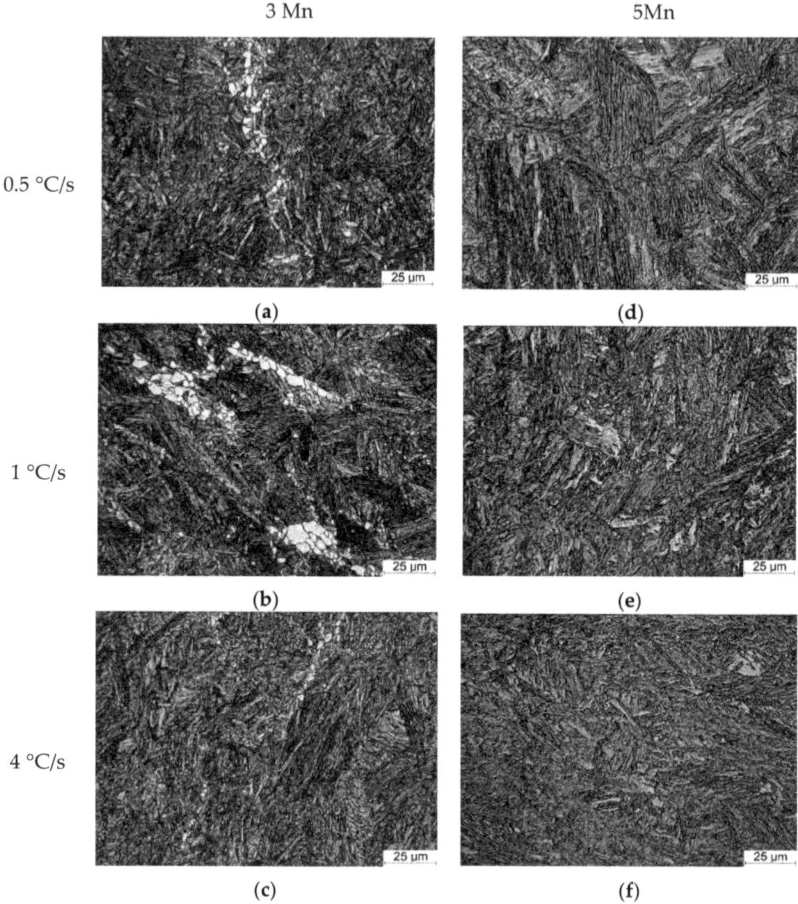

Figure 11. Microstructures obtained after deformation dilatometry tests for various cooling rates: (**a,d**) 0.5 °C/s; (**b,e**) 1 °C/s; (**c,f**) 4 °C/s.

Table 3. Cooling schedules following the multistep Gleeble simulations.

Temperature Range, °C	Cooling Rate, °C/s		
	3 Mn Steel Path 1	3 Mn Steel Path 2	5 Mn Steel Path 3
850 → 700	$V_1 = 30$		
700 → 650	$V_2 = 5$	$V_1 = 30$	$V_1 = 30$
650 → 450	$V_3 = 40$		
450	Isothermal holding at 450 °C within 300 s		
450 → RT	$V_4 = 0.5$	$V_2 = 0.5$	$V_2 = 0.5$
A Cooling Path			

Paths 2 and 3 cover simple heat treatments following the final deformation step at 850 °C. Taking into account that plastic deformation accelerates ferrite transformation [31] and that the accumulative strain after seven-step compression is larger compared to the single-step deformation (realized in the dilatometer), we also designed Path 1 (Table 3). The aim of this test was to check if the severe multistep deformation (compared to dilatometer conditions) will induce ferrite in the 3 Mn steel or not. Thus, the essential step of Schedule 1 was cooling in a 700–650 °C temperature range at a relatively slow rate (5 °C/s). This temperature range corresponds to the ferrite bay of the steel in the CCT diagram (Figure 7).

The LM and SEM micrographs after the isothermal heat treatments according to the cooling Paths 1–3 are shown in Figure 12. Firstly, it is obvious that the multistep deformation results in the advanced grain refinement of the microstructure when compared to the dilatometric specimens. It concerns both granular and lathlike structural components. The LM micrographs are not sensitive enough to distinguish individual phases due to the small size of the constituents. The only clear conclusion is that the microstructures do not contain polygonal ferrite (Figure 12a–c). The slow cooling according to the Path 1 did not induce ferrite, and the main structural constituent is carbide-free bainite of bimodal granular and lath morphology (Figure 12d). A lack of carbide leads to the enrichment of austenite in carbon, and finally its presence at room temperature. In Figure 12d this phase can be observed as bright granules (RA) embedded inside the bainitic constituents. The amount of retained austenite is smaller for the 3 Mn steel cooled directly from 850 °C (Figure 12e). The γ phase is located between laths of bainitic ferrite. One can see that only interlath-retained austenite is stable whereas some blocky grains are partially transformed into martensite forming martensite-austenite (MA) constituents. The steel containing the higher Mn content shows univocal lath morphology (Figure 12f). Retained austenite is located within bainitic-martensitic mixture, but the amount of the γ phase is only minor. Kamoutsi et al. [10] reported that a smaller fraction of retained austenite is caused by the smaller driving force for the enrichment of austenite in carbon with increasing Mn content. A more detailed quantitative analysis of retained austenite is reported elsewhere [36].

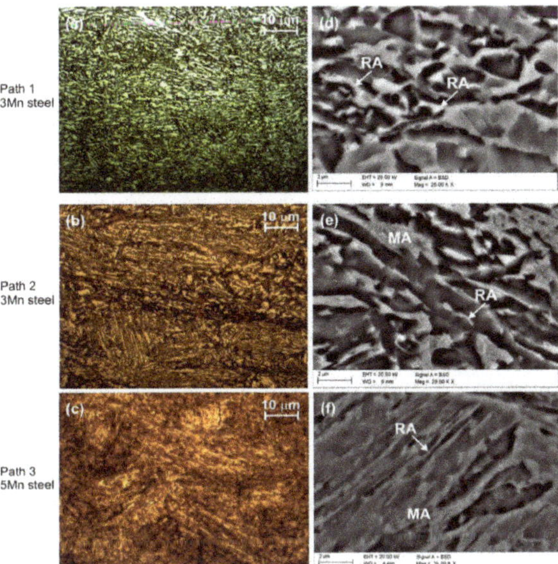

Figure 12. Multiphase bainite-based microstructures obtained after the thermomechanical treatment in the Gleeble simulator: (**a–c**) light microscopy (LM); (**d–f**) scanning electron microscopy (SEM); RA—retained austenite, MA—martensite-austenite constituents.

Morphological details of the microstructure have been assessed using transmission electron micrographs. Figure 13a shows a few granules of retained austenite within the granular bainite of relatively high dislocation density. The dislocations are a result of plastic deformation; however, some amount of them annihilates during slow cooling between 700 °C and 650 °C. The dislocation density increases for directly cooled samples because recovery processes are limited at 450 °C. Interlath-retained austenite of various thicknesses can be seen in Figure 13b. One can also see some austenite grains transformed partially into martensite (MA constituents). The 5 Mn steel contains predominantly continuous or interrupted laths of retained austenite embedded between the bainitic ferrite of high dislocation density.

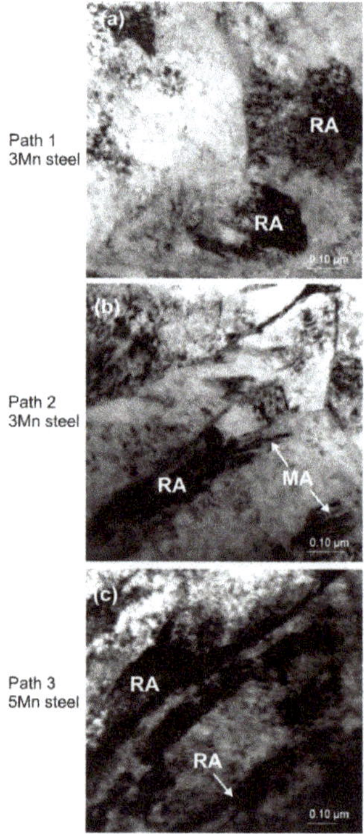

Figure 13. Transmission electron microscopy (TEM) micrographs showing the morphology of retained austenite and bainitic matrix in the: (**a**) 3 Mn steel (Path 1); (**b**) 3 Mn steel (Path 2); (**c**) 5 Mn steel (Path 3); RA—retained austenite, MA—martensite-austenite constituents.

The presence of possible MX phases (i.e., stable carbonitrides or carbides containing niobium) has not been revealed using a thin-film technique. Further experiments are in progress to confirm or negate this observation. However, our earlier investigations support the conclusion that there is a lack of MX carbides in the current medium-Mn steels. A reason for it is that the increase of Mn content increases the solubility of NbC and Nb(C,N) in the austenite and finally retards the precipitation of Nb(C,N) particles [36]. This effect is enhanced in steels containing a high content of Al fixing all of the nitrogen content.

4. Conclusions

The phase equilibrium state and austenite decomposition were analysed in two thermomechanically processed medium-Mn steels containing increased aluminium content. The increase of Mn concentration from 3% to 5% substantially affects the equilibrium conditions and austenite decomposition. Some discrepancies between the pseudobinary Fe-C diagrams and JMatPro calculations concerning a volume-fraction phase evolution were revealed. It especially concerns an interphase between the stable austenite and austenite + ferrite region during austenitization. These discrepancies are caused by untypical Al + Mn contents, which strongly affect a phase constitution. Real dilatometric tests confirmed a lack of ferrite in the steel containing 5% Mn and the occurrence of some small amounts of this phase in the 3 Mn steel.

The presented steels are prone to producing bainitic-martensitic mixtures during continuous cooling, even for relatively small cooling rates of 0.5 °C/s. The dominant structural constituent is bainite. The application of austempering allowed production of the fine-grained carbide-free bainitic mixtures with retained austenite. The higher fraction of interlath and blocky-type retained austenite was obtained in the 3 Mn steel, as a result of the hampering effect of Mn on the enrichment of austenite in carbon.

Acknowledgments: This work was financially supported with statutory funds of Faculty of Mechanical Engineering of Silesian University of Technology in 2016.

Author Contributions: A.G. conceived and designed the experiments and wrote the paper; W.Z. performed dilatometric experiments and analyzed the results; W.B. performed the equilibrium calculations; A.K. analyzed the data and reviewed the paper.

Conflicts of Interest: The authors declare no conflict of interest.

References

1. Gronostajski, Z.; Niechajowicz, A.; Polak, S. Prospects for the use of new-generation steels of the AHSS type for collision energy absorbing components. *Arch. Metall. Mater.* **2010**, *55*, 221–230.
2. Gibbs, P.J.; de Moor, E.; Merwin, M.J.; Clausen, B.; Speer, J.G.; Matlock, D.K. Austenite stability effects on tensile behavior of manganese-enriched-austenite transformation-induced plasticity steel. *Metall. Mater. Trans. A* **2011**, *42A*, 3691–3702. [CrossRef]
3. Rana, R.; Gibbs, P.J.; de Moor, E.; Speer, J.G.; Matlock, D.K. A composite modeling analysis of the deformation behavior of medium manganese steels. *Steel Res. Int.* **2015**, *86*, 1139–1150. [CrossRef]
4. Klancnik, G.; Medved, J.; Nagode, A.; Novak, G.; Petrovic, D.S. Influence of Mn on the solidification of Fe-Si-Al alloy for non-oriented electrical steel. *J. Therm. Anal. Calorim.* **2014**, *116*, 295–302. [CrossRef]
5. Wielgosz, E.; Kargul, T. Differential scanning calorimetry study of peritectic steel grades. *J. Therm. Anal. Calorim.* **2015**, *119*, 1547–1553. [CrossRef]
6. Grajcar, A.; Kuziak, R. Softening kinetics in Nb-microalloyed TRIP steels with increased Mn content. *Adv. Mater. Res.* **2011**, *314–316*, 119–122. [CrossRef]
7. Grajcar, A.; Zalecki, W.; Skrzypczyk, P.; Kilarski, A.; Kowalski, A.; Kołodziej, S. Dilatometric study of phase transformation in advanced high-strength bainitic steel. *J. Therm. Anal. Calorim.* **2014**, *118*, 739–748. [CrossRef]
8. Grajcar, A.; Radwanski, K.; Krzton, H. Microstructural analysis of a thermomechanically processed Si-Al TRIP steel characterized by EBSD and X-ray techniques. *Solid State Phenom.* **2013**, *203–204*, 34–37. [CrossRef]
9. Radwanski, K.; Wrozyna, A.; Kuziak, R. Role of the advanced microstructures characterization in modeling of mechanical properties of AHSS steels. *Mater. Sci. Eng. A* **2015**, *639*, 567–574. [CrossRef]
10. Kamoutsi, H.; Gioti, E.; Haidemenopoulos, G.N.; Cai, Z.; Ding, H. Kinetics of solute partitioning during intercritical annealing of a medium-Mn steel. *Metall. Mater. Trans. A* **2015**, *46*, 4841–4846. [CrossRef]
11. Sugimoto, K.; Tanino, H.; Kobayashi, J. Impact toughness of medium-Mn transformation-induced plasticity-aided steels. *Steel. Res. Int.* **2015**, *86*, 1151–1160. [CrossRef]
12. Lee, S.; Lee, K.; de Cooman, B.C. Observation of the TWIP + TRIP plasticity-enhancement mechanism in Al-added 6 wt. pct medium Mn steel. *Metall. Mater. Trans. A* **2015**, *46*, 2356–2363. [CrossRef]
13. Garcia-Mateo, C.; Sourmail, T.; Caballero, F.G.; Capdevila, C.; de Andrés, C.G. New approach for the bainite start temperature calculation in steels. *Mater. Sci. Technol.* **2005**, *21*, 934–940. [CrossRef]
14. Farahani, H.; Xu, W.; van der Zwaag, S. Prediction and validation of the austenite phase fraction upon intercritical annealing of medium Mn steels. *Metall. Mater. Trans. A* **2015**, *46A*, 4978–4985. [CrossRef]

15. Skolek, E.; Marciniak, S.; Świątnicki, W.A. Thermal stability of nanocrystalline structure in X37CrMoV5-1 steel. *Arch. Metall. Mater.* **2015**, *60*, 511–516.

16. Kokosza, A.; Pacyna, J. Formation of medium carbon TRIP steel microstructure during annealing in the intercritical temperature range. *Arch. Metall. Mater.* **2014**, *59*, 1017–1022. [CrossRef]

17. Jirkova, H.; Masek, B.; Wagner, M.F.X.; Langmajerova, D.; Kucerova, L.; Treml, R.; Kiener, D. Influence of metastable retained austenite on macro and micromechanical properties of steel processed by the Q&P process. *J. Alloy. Compd.* **2014**, *615*, S163–S168.

18. Mazancova, E.; Ruziak, I.; Schindler, I. Influence of rolling conditions and aging process on mechanical properties of high manganese steels. *Arch. Civ. Mech. Eng.* **2012**, *12*, 142–147. [CrossRef]

19. Steineder, K.; Schneider, R.; Krizan, D.; Beal, C.; Sommitsch, C. Comparative investigation of phase transformation behavior as a function of annealing temperature and cooling rate of two medium-Mn steels. *Steel Res. Int.* **2015**, *86*, 1179–1186. [CrossRef]

20. Mesquita, R.A.; Schneider, R.; Steineder, K.; Samek, L.; Arenholz, E. On the austenite stability of a new quality of twinning induced plasticity steel, exploring new ranges of Mn and C. *Metall. Mater. Trans A* **2013**, *44*, 4015–4019. [CrossRef]

21. Opiela, M. Thermomechanical treatment of Ti-Nb-V-B micro-alloyed steel forgings. *Mater. Tehnol.* **2014**, *48*, 587–591.

22. Lisiecki, A. Welding of thermomechanically rolled fine-grain steel by different types of lasers. *Arch. Metall. Mater.* **2014**, *59*, 1625–1631.

23. Bhadeshia, H.K.D.H. Anomalies in carbon concentration determinations from nanostructured bainite. *Mater. Sci. Technol.* **2015**, *31*, 758–763. [CrossRef]

24. Grajcar, A.; Lesz, S. Influence of Nb microaddition on a microstructure of low-alloyed steels with increased manganese content. *Mater. Sci. Forum* **2012**, *706–709*, 2124–2129. [CrossRef]

25. Kucerova, L.; Jirkova, H.; Masek, B. The effect of alloying on mechanical properties of advanced high strength steels. *Arch. Metall. Mater.* **2014**, *59*, 1189–1192.

26. Jablonska, M.; Smiglewicz, A. A study of mechanical properties of high manganese steels after different rolling conditions. *Metalurgija* **2015**, *54*, 619–622.

27. Zhao, L.; Top, K.A.; Rolin, V.; Sietsma, J.; Mertens, A.; Jacques, P.J.; van der Zwaag, S. Quantitative dilatometric analysis of intercritical annealing in a low-silicon TRIP steel. *J. Mater. Sci.* **2002**, *37*, 1585–1591. [CrossRef]

28. Thermo-Calc Software. Available online: http://www.thermocalc.com/products-services/databases/ (accessed on 3 October 2016).

29. Trzaska, J. Empirical formulae for the calculation of austenite supercooled transformation temperatures. *Arch. Metall. Mater.* **2015**, *60*, 181–185. [CrossRef]

30. Saunders, N.; Guo, Z.; Li, X.; Miodownik, A.P.; Schillé, J.P. Using JMatPro to model materials properties and behavior. *JOM* **2003**, *55*, 60–65. [CrossRef]

31. Grajcar, A.; Kuziak, R.; Zalecki, W. Third generation of AHSS with increased fraction of retained austenite for the automotive industry. *Arch. Civ. Mech. Eng.* **2012**, *12*, 334–341. [CrossRef]

32. Caballero, F.G.; Garcia-Mateo, C.; García de Andrés, C. Dilatometric study of reaustenitisation of high silicon bainitic steels: Decomposition of retained austenite. *Mater. Trans.* **2005**, *46*, 581–586. [CrossRef]

33. Bhadeshia, H.K.D.H. *Bainite in Steels. Transformations, Microstructure and Properties*, 2nd ed.; Institute of Materials, Minerals and Mining: London, UK, 2001; pp. 201–224.

34. Hase, K.; Garcia-Mateo, C.; Bhadeshia, H.K.D.H. Bainite formation influenced by large stress. *Mater. Sci. Technol.* **2004**, *20*, 1499–1505. [CrossRef]

35. Avishan, B.; Garcia-Mateo, C.; Morales-Rivas, L.; Yazdani, S.; Caballero, F.G. Strengthening and mechanical stability mechanisms in nanostructured bainite. *J. Mater. Sci.* **2013**, *48*, 6121–6132. [CrossRef]

36. Grajcar, A.; Skrzypczyk, P.; Kuziak, R.; Gołombek, K. Effect of finishing hot-working temperature on microstructure of thermomechanically processed Mn-Al multiphase steels. *Steel Res. Int.* **2014**, *85*, 1058–1069. [CrossRef]

MDPI
St. Alban-Anlage 66
4052 Basel
Switzerland
Tel. +41 61 683 77 34
Fax +41 61 302 89 18
www.mdpi.com

Metals Editorial Office
E-mail: metals@mdpi.com
www.mdpi.com/journal/metals